I0554810

Suburban Chickens

Raising Your Flock on Less Than One Acre

Karen Harris

LP Media Inc. Publishing
Text copyright © 2023 by LP Media Inc.
All rights reserved.
No part of this book may be reproduced or transmitted in any form or by any means,
electronic or mechanical, including photocopying, recording, or by an information storage
and retrieval system – except by a reviewer who may quote brief passages in a review to be
printed in a magazine or newspaper – without permission in writing from the publisher. For
information address LP Media Inc. Publishing, 1405 Kingsview Ln N, Plymouth, MN 55447
www.lpmedia.org

Publication Data

Karen Harris
Suburban Chickens - Raising Your Flock on Less Than 1 Acre – First edition.
Summary: "Successfully raising suburban chickens from chick to old age"
Provided by publisher.
ISBN: 978-1-954288-87-4
[1. Suburban Chickens - Raising Your Flock on Less Than 1 Acre – Non-Fiction] I. Title.

This book has been written with the published intent to provide accurate and
authoritative information in regard to the subject matter included. While every reasonable
precaution has been taken in preparation of this book the author and publisher expressly
disclaim responsibility for any errors, omissions, or adverse effects arising from the
use or application of the information contained inside. The techniques and suggestions
are to be used at the reader's discretion and are not to be considered a substitute for
professional veterinary care. If you suspect a medical problem with your chickens, consult
your veterinarian.

Design by Sorin Rădulescu
First paperback edition, 2023

TABLE OF CONTENTS

Chapter 6

Chapter 7

Chapter 8

Chapter 9

Chapter 13

Health and Wellness of Your Flock 207

Chapter 14

Sustainable Practices for Suburban Chicken Farming 227

Chapter 15

CHAPTER 1

Introduction to Suburban Chicken Farming

The Charm of Chickens

> *Talk to your chickens, enjoy their personalities, and have fun! At the beginning it was about the eggs ... now it's about my girls. I have an 11-year-old chicken! In the summer I'll drink my coffee out back and enjoy watching them as they run around. They recognize our voices and come running when we are out back with them.*
>
> AMY JOHNSON

A quaint country scene of plump hens pecking and scratching the ground as an adorable child, squealing with the thrill of discovery, gathers fresh eggs in a basket for the morning's breakfast ... that's the romantic vision many people have of the rural lifestyle. There is something very nostalgic about tending a flock of chickens. It reminds us of our grandparents. It reminds us of Laura on *Little House on the Prairie*. It reminds us of our agrarian roots.

I will admit it. That's one of the reasons why I ordered my first batch of chicks nearly 25 years ago. Sure, it's nice to have a ready supply of eggs, but it was more than that. The kids were young, and I wanted to teach them responsibility. I wanted them to have an understanding of where

our food comes from and a respect for the farmers who feed our nation. Raising chickens in our backyard checked all those boxes—and more.

When the kids showed chickens in 4-H, they learned about competition and—about 75% of the time, in our experience—how to lose graciously. Although I didn't know it when I placed the order for our first chicks, raising chickens also taught our kids some tough lessons about the circle of life ... how hens don't have a particularly long life span, and how, unfortunately, chickens can fall prey to predators. A flock of hens can be cliquey. There's a definite social hierarchy—a pecking order—and hens on the lower rungs are targets of bullying. When my daughters were in their tumultuous middle school years, I was able to draw some analogies between the different personalities of our chickens and the mean-girl cliques to help them see the destructive power of pecking orders.

Chickens are the easiest of all livestock animals, as well as being the cheapest to raise. The argument could be made that you get the highest return on investment with chickens. And chickens are useful too. We can make good use of their meat, eggs, feathers, and guano.

Suburban America

Fortunately, chickens are one of the most adaptable livestock animals, too, which makes them ideal for suburban life. Chickens don't require large barns or expansive pastures. If you have a small patch of backyard, you can raise a small flock of happy, productive chickens.

When you look at the history of the United States, suburbs are a relatively modern invention. In the prosperous post-World War II era, more families could afford to move out of the cities—but not too far out. And they could afford automobiles so the family breadwinner could commute to a job in the city. Suburbs provided an opportunity for families to own a small piece of soil and raise their children in a setting that was not as crowded, hot, and dirty as an urban community and also not as remote and dirty as a rural one. Suburbs remain a happy medium.

For a couple raising a family, suburban living offers a neighborhood feel, other children to play with, relative safety, and better schools. Some suburbs, however, have some drawbacks. If they are technically part of the larger city they surround, the suburbs may lack political autonomy and have little voice or representation in the governing body that oversees them. As we will see in later chapters, this can sometimes be a detriment to suburban chicken farmers. You may own your piece of suburban land, but that doesn't necessarily mean you are free to do with it as you want.

How Suburban Chicken Farming Differs from Traditional Chicken Farming

Raising chickens in a suburban environment differs from traditional chicken farming. A chicken farmer in a rural setting can maintain a larger flock with upward of 100 chickens. He or she may have multiple roosters, though, as we will see later, having too many roosters can be problematic. The chickens might free-range during the day and come home to roost in a sizable coop in the evening. As Kristen Harrell, a suburban chicken farmer living in Iowa, explains, "My personal philosophy is that my chickens are happier and healthier with the opportunity to free range around my yard than they would be if they were locked up."

For suburban chicken farmers, the flock will be much smaller—maybe only four or five hens and no roosters. The birds may be kept in a fenced enclosure or chicken run and roost in a small coop, chicken tractor, chicken ark, or other type of portable poultry housing.

Beyond local ordinances, housing type, and flock size, however, there is little difference between raising chickens in a suburban setting versus a rural one. Hens only require a healthy diet, clean water, fresh air, protection from predators, and plenty of sunlight to do what they do best—lay eggs.

Photo Courtesy of Kristen Harrell

Benefits of Raising Suburban Chickens

> **"**
>
> *Being a chicken owner is far more rewarding than I had expected. I love having fresh eggs, but I also enjoy seeing the personality of each hen. It is a wonderful system to be part of. The chickens eat weeds and bugs, produce eggs, and provide fertilizer. I wish that every neighborhood would have a little flock where people could drop off food scraps and access fresh eggs.*
>
> KRISTEN HARRELL
>
> **"**

Farm Fresh Eggs

As I write this, increasing food prices are a national concern. Every day, there is at least one news article on the rising costs of eggs. For some reason, a dozen eggs are the indicator that economists like to use to

Photo Courtesy of Anne Greenwood

show the impact of inflation at the grocery store. I am not sure why this is, but my guess is that it is because eggs are a staple. Nearly everyone buys eggs. The current news stories are tossing out statistics about egg prices, noting that they have gone up as much as 50%, 60%, and even 65% in the last year or so.

If your family are big egg eaters—and they should be! Eggs are a wonder food—you may have felt the pinch of higher egg prices. A small backyard flock of hens, once they reach egg-laying maturity, will keep you well supplied with eggs. You may even get enough eggs to sell to your friends and neighbors.

The eggs that will come from your own chickens will taste better than the eggs you buy at the supermarket, partly because of the quality and variety of food you will feed the hens and partly because of freshness. Commercial egg farmers, by law, have up to 30 days to package their eggs and ship them to market. Stores, by law, can then keep eggs on the shelves for an additional 30 days. That carton of eggs that you buy could be up to two months old by the time it makes it into your refrigerator. The quality and taste are affected. You will be surprised at how delicious fresh eggs are.

FUN FACT

Ancient Companions

According to archaeological evidence, the origin of the backyard chicken may date back over 7,000 years. The earliest fossilized chicken bones come from northeastern China and can be dated to around 5400 BCE. Some experts believe chickens were the first animal to be domesticated by humans.

Sustainable Food

These days, we are all more aware of the impact humans are making on the environment. Believe it or not, raising suburban chickens is a great first step toward living a more sustainable lifestyle. You will be producing your own eggs and, if you so desire, your own meat. Additionally, chickens will help control the insect population in your yard without the need for harsh chemical pesticides. The soiled bedding from the chicken coop can be added to your compost pile. Chicken poop is rich in nitrates, so it is an ideal fertilizer for your vegetable garden or flower beds. You will have plump tomatoes and verdant peppers to serve with your omelets.

A Rewarding Hobby

Raising chickens is an enjoyable and rewarding hobby for families with young children, singles, and empty nesters. Chickens are funny, gregarious, personable birds, and they truly seem to enjoy interacting with us as much as we enjoy them. It is not uncommon for me to find my husband sitting in a lawn chair next to the chicken coop after a hard day at work. Watching hens strutting around, chasing bugs, and pecking at the ground is somehow very therapeutic and relaxing. It's the farmer's de-stressor.

According to Amy Johnson, who raises suburban chickens at her Minnesota home, "In the summer, I'll drink my coffee out back and enjoy watching my chickens as they run around. In the beginning, it was all about the eggs, but now it's about my girls." She suggests, "Talk to your chickens, enjoy their personalities, and have fun!"

Educational Opportunities

I know from my own experience that raising backyard chickens helped my kids develop an excellent work ethic and learn about responsibility. They felt proud when they dropped off a dozen eggs to a neighbor and brought a few baby chicks to school for show-and-tell. Our kids named every chicken we've ever had (hundreds of them!), and we still reminisce and tell stories about our favorite hens—and our least favorite roosters!

Overview of Challenges and Considerations

Local Ordinances

Keeping chickens can be a positive experience, but there are some challenges that are unique to suburban chicken farming. You may run into conflicts with local ordinances and neighborhood association rules, although more places are now willing to make accommodations for small suburban chicken flocks. Later in this book, we will discuss how to find out what the laws are in your area and, if necessary, how to present current, up-to-date information about small-scale chicken farming to lawmakers in hopes of changing the ordinances.

Photo Courtesy of Anne Greenwood

Housing Considerations

In a suburban setting, the housing you select for your chicken flock is important. It needs to meet all the needs of the chickens, which we will outline in a later chapter, but it also needs to be aesthetically pleasing so your neighbors or your homeowners' association doesn't complain of an eyesore. Lastly, you need to have a plan for housing your hens when they are tiny chicks, as well as during the cold winter months.

Time Commitment

You need to make sure that chickens fit into your lifestyle. If you are a busy professional who travels a lot for work, you probably don't have the time to tend a suburban flock. If you have a busy family and your evenings and weekends are filled with soccer practices, dance lessons, art club meetings, and piano recitals, you need to ask yourself if you can add chickens to the mix. Chickens don't require much of your time, but they do need tending every day.

Being Considerate of Your Neighbors

Suburban chicken farmers must consider their neighbors, something rural chicken farmers usually don't have to think about. As we will discuss later, you, as a responsible chicken farmer, need to make sure that your hens are not causing problems for your neighbors. We will talk about the noise and the smell of the birds, as well as potential health risks. Your neighbors may be concerned about avian flu, so you should have some answers for them.

Protection from Predators

Both rural and suburban chicken farmers need to protect their flocks from predators, but you may find that there are different types of predators in a suburban setting. A suburban chicken farmer is more likely to have trouble from neighborhood dogs than coyotes, but the same precautions are necessary. You need a coop and pen that will keep your chickens in and predators out.

Diseases

In a subsequent chapter, we will learn ways to decrease the risk of disease sweeping through your suburban flock. Since diseases can be transmitted to domesticated birds from wild birds, you need to know how to keep your hens away from Canada geese, ducks, pigeons, and other birds.

Chapter 1 Summary

In the quarter of a century that I have raised chickens, I have found that the challenges have been greatly offset by the benefits. Chickens are fairly low maintenance, but they aren't maintenance-free. I only spend about 10 to 15 minutes per day attending to my chickens' needs and about an hour or so on the weekend cleaning out their coop. In return, I get, on average, one egg per day from each of the hens throughout the spring, summer, and fall. Most hens, as we will see later in this book, slow down on egg production in the winter.

I am admittedly biased, but I don't have too many negative things to say about raising chickens in a suburban setting. Overall, it has been a wonderful hobby. Everyone in the family has done their fair share of chicken chores, and we have been able to bond as a family over our small flock. And I am not the only one that feels this way. Kristen Harrell of Iowa explains, "Being a chicken owner is far more rewarding than I had expected. I love having fresh eggs, but I also enjoy seeing the personality of each hen. It is a wonderful system … the chickens eat weeds and bugs and produce eggs and fertilizer."

If you are ready to start your adventures in suburban chicken farming, you have come to the right place. Not only will we discuss housing, fencing, and feed requirements in this book, but we will also have chapters covering chicken breeds, where to purchase your chicks, and the pros and cons of keeping a rooster.

CHAPTER 2

Sourcing Chicks

O nce you have made the decision to start your suburban chicken flock, you need to decide what type of chickens to get, what breed you want, when to get baby chicks, and where. This may seem like a lot of decisions to make, but this chapter will help you answer those questions and give you tips for preparing for the arrival of your chicks.

Chicken Terms

Before we begin to answer these questions, a quick lesson in chicken vocabulary is in order. These are terms that you will hear or read when you are in the market for baby chicks, so it will be helpful to know what they mean.

Chickens by Age:

Chick – From the time they are hatched until they are about three months old, baby chickens (both males and females) are called chicks.

Pullet – A female chicken over 12 weeks in age and under one year in age is called a pullet.

Cockerel – A male chicken over 12 weeks in age and under one year in age is called a cockerel.

Hen – A female chicken over the age of one year is called a hen.

Cock/Rooster – A male chicken over the age of one year is called a cock or a rooster.

Chickens by Variety:

- **Bantam** – A bantam is a variety of chicken that is much smaller than regular chickens. Breeds of bantams can be quite unique and ornamental, which makes them ideal show birds. Some bantams are good egg layers. Bantam eggs are just as nutritious and tasty as large-breed chicken eggs; however, they are smaller in size. There are hundreds of ornamental bantam breeds.
- **Large Breed** – A large-breed chicken is probably the type of chicken you are picturing in your head. Most stand between 13 and 18 inches tall. The American Poultry Association recognizes 65 varieties of large-breed chickens.
- **Layer** – A chicken breed that is best suited for egg production.
- **Meat Bird** – A chicken breed that is best suited for meat production.
- **Dual Purpose** – A chicken breed that is a good egg producer and plump enough for meat production.

Chicks by Gender:

- **Straight Run** – A group of baby chicks that are a mix of males and females.
- **Sexed** – A group of baby chicks that have been separated by gender. It usually means a group of female chicks.

Layers Versus Meat Birds

Now that your vocabulary lesson is over let's tackle your first decision: What type and breed of chicks should you get?

If you are like most suburban chicken farmers, you are in it for the eggs. I will tell you from firsthand experience that raising meat chickens to put in your freezer is a lot messier and more unpleasant than it sounds.

I will digress for a moment to tell you that we did, for a short time, try our hands at raising meat birds. I would not call the experience a success. For starters, the kids were freaked out. There were tears. Second, my husband, who was certain that he could carry out the hatchet duties like a macho pioneer man, found out he was a bigger softie than he cares to admit. And I was tasked with plucking the chicken. I was pregnant at the time and living in nausea land. Plucking chickens pushed me over the edge. Lastly, the kids refused to eat any meals made with chicken until they were certain all the packages were gone from the freezer.

Don't let my experience sway you, though. It is possible to raise meat birds in a suburban setting. I would just suggest that you refrain from DIY butchering and take your birds to a professional when the time comes. We actually did this a few times. We simply dropped off a few of our live chickens in a large cat carrier and returned to the butcher shop a few days later to pick up the neatly packaged, already-frozen meat. The cost was surprisingly low (at that time), and it saved us from getting our hands dirty.

Have a discussion with your family about your hopes for your back-yard chickens. Look at breeds that are good layers if your focus is on egg production. Look for meat birds if you want to go that route. Maybe a dual-purpose chicken breed will suit your purposes best. If you want chickens more as pets or if you think you might want to enter some chicken shows, perhaps you should think about bantams.

Chicken Breeds

> **"**
>
> *Any type of breed will work. Certain breeds are better egg layers. I prefer Ameraucana or ISA Brown, as they are great layers. My kids love Polish Crested because they are smaller and the kids can carry them around and play with them. Silkies, Ayam Cemani, and Turkens are others we have enjoyed.*
>
> AMY JOHNSON
>
> **"**

In the next section, we will offer some suggestions on chicken breeds to consider. These breeds were selected not only because they will accomplish what you want them to accomplish—like lay a bunch of eggs—but because they are, in general, docile, non-aggressive, quiet, non-broody breeds. When selecting a chicken variety, you should always do your homework. Don't just pick a chicken breed based on its appearance, but research its temperament, characteristics, and special considerations.

Layers

Rhode Island Reds

Barred Rocks

White Leghorns

Buff Orpingtons

Americanas

Wyandottes

Meat Birds

Jersey Giants

Bresse

Cornish Crosses

Brahmas

Dual-purpose Chickens

| Delawares | Black Australorps | Speckled Sussex |

Bantams

| Sebright | Old English Game | Frizzle | Silkie |

"Personally, I have had great luck with Delawares and Barred Rocks." Elizabeth Sorby of Washington shares, "Americanas are great all-purpose birds that lay blue eggs and are reliably good pets. Every Americana I've had is the friendliest hen in my flock."

Another Washington resident, William Quigley, adds, "I have never had a 'bad breed,' except for one Blue Wyandotte that had a bad temper. Leghorns seem to lay the biggest eggs."

When to Get Your Chicks

For the best chances of success with your new flock of chickens, timing is everything. You may decide in September that you want to get chickens, only to find that your local farm and feed store no longer has a metal trough full of baby chicks like it did in the spring. It is possible to get chicks year-round, usually from a local hatchery or by mail order, but most chicken farmers get their chicks in early spring, about the time you see those metal troughs pop up at the farm and feed stores.

The reason for this has more to do with tradition than with good farming practices. Wild chickens, like most animals, typically lay eggs and hatch chicks in the early spring; therefore, early chicken farmers followed suit. Today, chicken hatcheries operate throughout the year, so if you want to start your flock in September, you can find chicks.

Where to Get Your Chicks

There are several ways to obtain baby chicks to start your suburban flock. Naturally, there are pros and cons to each of them. You could visit a local hatchery, mail order your chicks from a non-local hatchery, purchase them at a local feed store, find a breeder in your area, or look at ads in the newspaper or on social media market pages. Let's look at the pros and cons of each.

Local Hatchery

I have purchased chicks from a local hatchery. Well, it was about an hour's drive away. I had to take time off work because they only allowed the public in to purchase chicks from 8:00 a.m. to noon on Wednesdays. I went in the early spring; it was the first week of March if I recall. As I got close to the hatchery, it started to snow. I live in a lake-effect area, so within minutes, the snowfall was quite heavy.

I picked up the chicks, placed the box on the passenger seat, cranked up the heat for them, and then started the drive home. On the highway, I suddenly felt the car dip and pull. I had a flat tire! It was still really snowing, and visibility was poor, so I pulled as far off the road as I could and fretted about what to do. Here I was, stranded on a busy highway with a box full of freezing chicks.

Then I heard a knock on my window. It was a state police officer stopping to see why I was pulled over. I explained the situation and added

Pros

- A local hatchery is close to you. You can stop there, see the facility, and bring your chicks home right away if they are available.

Cons

- Local hatcheries are often limited in the varieties of chickens they have. If you are looking for a specific breed, your nearest hatchery may not offer it. Also, chicks aren't always available. You'll have to do your homework ahead of time to see when they will have chicks for purchase.

that I was worried about my new chicks. He seemed a bit surprised, then laughed as I opened the box for him. Even though the weather was terrible, he offered to change my tire for me so I wouldn't have to shut the car off and risk the chicks getting cold. Great guy!

Mail Order

Pros

- ⊘ You can get just about any variety of chicken through mail order. You can get chicks delivered any time of the year. The chicks are sexed, so you can order only hens if you want.

Cons

- ⊗ You won't get your chicks immediately. There is a slight chance that one or more of the chicks will not survive the shipping process.

There are several large, well-established chicken hatcheries located around the country that offer a variety of chicken breeds. If you are looking for a specific breed of chicken that is not common in your area, you will be able to find it at one of these hatcheries.

You can look through a catalog of chicken breeds or peruse the hatchery's website to see colorful photos or illustrations of the different chicken breeds. They even include baby pics of the various breeds so that you can see what the chicks will look like when you get them versus how they will look when they are fully grown. You will find helpful information about each breed as well, which can assist you when selecting the right breed.

Feed Stores

For chicken lovers, there is a big rush of excitement when the local feed stores get their shipments of baby chicks in the spring. The stores set up a cluster of silver metal feed troughs with heat lamps clamped to the sides and fill them with sawdust and chicks. Customers love this, especially kids. That is the root of some of the problems with feed store chicks. People will pick up one of the cute, fluffy chicks, then put it back in the wrong trough.

ORDERING CHICKS THROUGH THE POSTAL SERVICE

When people, especially ones who are new to chicken raising, find out that hatcheries mail baby chicks through the post office, it understandably raises a few questions—and eyebrows. Stuffing newly hatched chicks into a box, slapping a few stamps on it, and entrusting it to the United States Post Office does sound bizarre and cruel. Let's be honest; the USPS has a reputation for less-than-speedy deliveries and even lost parcels. That might be true of the socks you ordered or the postcard your mother sent you from their vacation to the Alamo, but with chicks, the post office's reputation is first class. Hatcheries have been sending baby chicks this way for more than half a century, so they have the system down to a science. And so does the post office.

Hatcheries are businesses. They want their customers to be happy and satisfied. They want their product—the chicks—to arrive safely at their destination. The hatcheries wouldn't utilize the post office if they didn't think it was an efficient and timely delivery service.

Now, let's take a quick moment for a biology lesson. You know how the inside of an egg has a white part and a yellow yolk part? The baby chick actually forms from the white part, the albumen. The yellow part, the yolk, is a food source for the chick. It is packed with liquid nutrients that the chick absorbs right before it hatches. The yolk provides all the food and water the chick will need for its first 72 hours of life. They don't need to eat or drink for the first three days. This is an adaptation that chickens developed in the wild when not every egg hatched at the same time. For domestic chickens, this adaptation gives the hatchery and the post office 72 hours to get the newly hatched chick into your hands.

The folks working at large hatcheries have the chicks hatching on a tight schedule. The eggs are incubated at the same time so that they all hatch at once, usually within a carefully controlled window of time. As soon as the chicks emerge from their shells, the hatchery workers sort them and fill orders. The chicks are packed into boxes that are small enough that the chicks are forced to huddle together; that keeps them warm. The labels and postage are added, making sure the breathing holes in the boxes aren't covered, and the shipment is whisked off to the nearest post office, where the postal workers are awaiting its arrival.

At the post office, the boxes are scanned and sorted. Many of them end up on planes, flying to various parts of the country. Others are put on trucks and sent to the next post office. The boxes of chicks remain in transit around the clock for the next day until they make it to local post offices early in the morning on the third day, usually by 5:00 a.m.

In most cases, the post office will call customers to let them know their chicks can be picked up at the post office. In rare cases, the mail carrier will actually deliver them to a customer's home. Every time I've mail-ordered chicks, I've gotten an early morning phone call from the carrier at my hometown post office. I drive there in my pajama pants to pick up my box of peeping babies.

Depending on the time of year and the temperature at the destination, the hatchery may pack the chicks in an insulated box or include a small heat pack that looks like one of those hand-warmer things. They may even add a Gro Gel—a packet that provides nutrients and hydration for the chicks—to the box, just in case there is a delay.

As soon as customers get their mail-order chicks home, the birds should go right under a heat lamp. Take each chick individually and dip its beak into water. Remember, the chicks have yet to eat or drink anything. They don't know how. Introduce each one to the waterer and the feeder. They will figure it out from there.

Also, count your chicks. Most likely, you got an extra one or two. Hatcheries do this in case a chick or two doesn't survive the mailing process. If any of the chicks are dead, be sure to contact the hatchery right away. They will replace the fallen chick. I have purchased chicks through the mail dozens of times, and I can only think of about three times when I opened the box to find a dead bird. I have never had more than one dead chick per order.

When I explain to friends and neighbors that our new chicks came by mail, they are often shocked. I have to admit that it does sound barbaric, but it is a process that has been going on successfully for decades. If that doesn't convince you to give mail-order chicks a chance, let me add one more point. All those baby chicks that are available for sale at your local feed store were shipped to that store through the mail. That's how feed stores get their chicks too.

You can see how easy it is for the chicks to get all mixed up. I have ended up with unwanted roosters this way. Someone picked up chicks from the "straight run" trough and put them back in the "pullet" trough. I'd like to think this was done by accident, but it could be that a prankster thought it was funny to mess with someone's future flock.

With so many people handling the baby chicks in feed stores, diseases can spread quickly. I personally haven't purchased any feed store chicks that turned out to be sick, but I have heard of it happening.

Pros

⊘ Baby chicks are often cheaper at feed stores. In the spring, they are readily available.

Cons

⊗ Feed stores don't usually carry a wide variety of chicken breeds. The chicks get mixed up, so you are never sure you are getting what you want. The chicks may have diseases.

Local Breeders

In my area, most of the local chicken breeders produce specialty birds for 4-H and poultry shows, not your average egg layers for a backyard flock. But it is worth investigating your area to see what breeders are there and what types of chickens they produce. When you source your chicks locally, they will likely be healthier birds. But be careful too. Some small-scale breeders aren't as careful about responsible breeding practices as they should be. You may end up with some crossbreed chicks that don't have the qualities you are looking for.

Pros

⊘ Local breeders are usually small-scale chicken farms that are dedicated to producing healthy, quality birds. The local breeder can answer many of your questions and direct you to resources in your community.

Cons

⊗ Local breeders usually offer only a few breeds. The cost is typically higher.

Ad in a Local Newspaper, Flyer at a Feed Store, or a Facebook Community Exchange Group

Pros

⊘ About the only pro to this would be that your chicks will be local. And cheap.

Cons

⊗ Chances are, if someone has posted an ad selling baby chicks, they are males who will grow up to be roosters. That's why the person is selling them. They could also have been exposed to a disease, and the current owner doesn't want to destroy them but also doesn't want to risk infecting their mature chickens.

I'm on a few of those Facebook community marketplace groups, and from time to time, someone posts about having chicks available. Many times, there are posts saying the chickens are free because the owner can no longer have them. I would question this. Is the owner getting rid of the birds because they ended up with too many roosters? Are the birds ill? How old are the birds? There are so many other viable sources for getting your chickens that you don't need to get them from a random stranger on Facebook.

With that said, I am not against adopting hens from people I know. The members of the community garden in the largest city near me, for example, always get a few chicks in the spring and keep them in a mobile coop at the community garden. They are a nice addition to the community garden because they supply eggs to the members, eat the insects, and provide a ready source of fertilizer. But the community garden folks are not equipped to house the hens over the winter. Every fall, a couple of hens come to live at my henhouse until the weather gets warmer and they can move back to the city.

Factors to Consider When Selecting Chicks

Earlier in this chapter, we discussed the types of chickens, like layers and meat birds, as well as the breeds of chickens. These are, of course, important factors when selecting chicks, but there are a few others, particularly the age and sex of the chicks.

Age

If you recall our vocabulary lesson, you will know that chicks are chicks until they reach about 12 weeks of age, and then they are pullets and cockerels. You will most likely get young chicks if you purchase them from a feed store or mail-order them from a hatchery, but it is sometimes possible to find pullets for sale. That will speed up the time it takes for them to start laying, and you won't have to worry about housing

young chicks until they are old enough to be in the coop, a topic we will cover next.

Gender

Most suburban chicken farmers don't want—or can't have—a rooster. In a later chapter, we will go into depth about the pros and cons of keeping a rooster, but for now, we will focus on hens and pullets.

When you purchase chicks as a straight run, that means you are getting a mix of males and females. In theory, it should be a 50-50 mix, but that hasn't been my experience. I usually end up with more males. Why? Because the hatcheries also sell sexed chicks or all female chicks. When they pull out the female chicks, it skews the numbers.

Even when you order sexed chicks, you are not always guaranteed to get only females. Sexing newly hatched baby chicks is a difficult thing to do. It takes an expert with a keen eye to do this accurately. A hatchery employee spreads open the chick's vent to observe the shape of the organs. It is not as easy as determining the gender of a kitten or puppy. There are 18 different shapes that might be observed, so the hatchery worker needs to know which ones indicate male and which ones indicate female. On top of that, some of the shapes closely resemble others. With training and experience, a hatchery chicken sexer has between a 90% to 95% success rate.

Once the chicks get a little older, determining the gender is easier. Sometimes the feather colors are different between pullets and cockerels. You can also observe the bird's comb. Cockerels usually have a more prominent comb.

FUN FACT

Chicken Personalities

Like humans, chickens can display individual personalities and a range of behaviors. For example, some chickens are outgoing, friendly, and curious, while others are stand-offish or independent. Because some chicken personality traits are linked to breed, it's crucial to research chicken breeds before purchasing your flock to ensure the best possible outcome.

Preparing for Your Chicks' Arrival

Before you impulse-buy some baby chicks at your local feed store, you need to have things prepared at your home. You can't just toss the chicks into your coop. They are just babies. They need a bit of extra care for the first few weeks. Let's go over some tips to help you get everything ready for your chicks' arrival.

Select an Appropriate Place for the Brooder

As I said, you can't immediately house your chicks in your outdoor chicken coop. They are too young and fragile at this point. The cute, fluffy down covering their bodies is not the best insulator. Once their feathers come in, they will be better equipped to handle cooler temperatures, especially at night. Until then, set up a small brooder for them in your

house or garage. Be sure the place you choose is secure, warm, and well-ventilated. I set my brooder up in our mudroom, so it is out of the way of normal household activities, yet in a heated room.

Set Up a Brooder

Suburban chicken farmers raising four to six chicks don't need a large or elaborate brooder. You can use a cardboard box—lots of people do—but be sure to line it with plastic so the cardboard doesn't become soaked. A plastic storage tote also works well, as does a plastic kiddie wading pool. Look for something with high enough sides to keep the chicks contained. Believe me, as they get older, they will learn to jump out.

Line the bottom of the brooder box with some absorbent material, like sawdust, wood shavings, or shredded newspaper. You will want to change the bedding on a regular basis. It gets stinky pretty quickly.

Provide Heat, Light, and Ventilation

Baby chicks need warmth, so you will want to set up a heat lamp. Be careful of how and where you set it up. You don't want to risk starting a fire. As the chicks get older, they will get more rambunctious and will jump on top of anything in their brooder, including the heat lamp. They may even knock it over, and since you have lined your brooder with wood shavings, sawdust, or newspaper, this can be a recipe for disaster. I hang my heat lamp and suspend it above the brooder. This gives me some peace of mind and also allows me to move the heat lamp up and down as necessary.

Newly hatched chicks need the temperature in their brooder to be about 90 to 95°F for the first week or so. Make sure there is room for them to huddle under the heat lamp when they are cold and get out from under it if they need to cool down. After the first week, you can begin to gradually decrease the temperature every few days to help the chicks acclimate to cooler weather.

The heat lamp provides all the light that the chicks need, but they also must eventually learn that darkness exists. When they get a little older, and they are not in danger of getting chilled, shut the heat lamp off for an hour or so. The chicks will freak out at first—lots of peeping, flapping, and running about—but it is better for them to experience darkness at a young age before they move out to the coop.

Gather Supplies and Equipment

You need to supply your baby chicks with plenty of food and water, so you will need a chick waterer and a feeder. You can find these items at your local feed store or order them from Amazon. They are inexpensive. The first thing you will probably notice is that the feeder and the part of the waterer that holds the water will be red. All of them are red. Chickens have remarkable eyesight and can see colors, unlike many animals. They are attracted to the color red, which is why manufacturers make waterers and feeders red. It is also why chickens will decimate your tomatoes if given the chance.

Have plenty of chick starter—food for the chicks, which we will discuss in greater detail in the next chapter—on hand so you can keep the feeder full.

Quarantine New Chicks

Diseases spread quickly between birds, so you will need to take precautions with your chicks. If this is your first time getting chickens, you won't have to worry so much about quarantining them, but if you are adding new chicks to your existing flock, you will. The standard time frame for quarantine is 30 days. During that time, do not allow your chicks to come into contact with your older hens. Do not use the same waterers or feeders, and be sure to thoroughly wash up and change your clothes when you go from tending one group to the next. For new chicks, try to keep them away from wild birds as much as you can, though this can sometimes be a challenge. And, of course, no chick playdates with the neighbor's chicks—at least, not for the first month.

Socialize Your Chicks

Your new chickens will be part of your family, and you can start right away to teach them to trust you and your family. Pick up the chicks and hold them. Stroke their feathers. Allow them to eat out of your hand. Talk to them; even sing to them! The more your chicks get used to being handled when they are young, the friendlier they will be as full-grown hens.

Photo Courtesy of Anne Greenwood

Just remember that baby chicks are delicate. If you have small children at home, remind them about this. Scoop up the chick from the underside and hold it in the palm of your hand. Place your other hand loosely over the top of the chick. These little birds can be quick and impulsive. One may jump from your hand. That's a long way down for such a little guy!

When you are done holding a chick, set it down gently. Always, always supervise your children when they are playing with the chicks. Remind your kids to be gentle and quiet and to avoid sudden movements that will startle the chicks. Beyond that, have fun getting to know your chicks. Name them. Learn their personalities. Marvel at how quickly they grow and change. Soon, they will be ready to move to their permanent quarters.

Chapter 2 Summary

This chapter covered a lot of information, but all of it is important to consider before you start on your journey as a suburban chicken farmer. You need to decide what type of chicks to get, where to get them, and how to prepare for their homecoming. Once you make your decisions and find your chicks, the real excitement can begin.

CHAPTER 3

Raising Chicks

While the last chapter covered ways to prepare for chicks and places to buy your chicks, this chapter will focus on raising your chicks from the time you first bring them home to their brooder box until they are ready to move outside to their coop. The tips outlined in this chapter will help you give your new flock a great start toward becoming happy, healthy, and productive hens.

Feeding and Watering Your Chicks

Chicks grow tremendously fast in the first few weeks of life, which is why proper nutrition is vital. Your chicks should have constant access to fresh, clean water and be fed on a regular basis.

A Lesson in Eating and Drinking

We touched on this briefly in the last chapter, but it is worth repeating. If you have acquired your chicks from an out-of-town hatchery, and they were shipped to you through the postal service, they have never had an opportunity to eat or drink. They need some instruction. When you open the box at home, remove the chicks one by one, inspect them, and dip their beaks into the water. The chick will act surprised and shake its head, but it will quickly figure out how to find and drink the water. Do the same thing with the feed. Carry the chick to the feeder and put its beak into the feed.

I would strongly suggest doing this one at a time as you take the chicks out of the box. Chicks are fast, and they all look alike. If you put

Photo Courtesy of Casey McMahon

them into the brooder box first, you may miss one of the chicks when you show them the water and feed. I would also suggest that you spend a few moments watching them to make sure that they are all active and healthy. If you notice one that isn't eating or drinking, give it another lesson. They are quick learners; plus, they are probably hungry. They will figure out where their food and water are in no time.

Chick Starter

For the first few months, you will feed your chickens chick starter, an umbrella term for feed that has been specially formulated to meet the nutritional requirements of growing chicks. Here is what you need to know about chick starter and what to look for when making your feed purchase.

Photo Courtesy of Kristen Harrell

Chick starter is made primarily of grains that have been crushed into a fine mash. The tiny chunks are the ideal size for young chicks to pick up with their beaks and digest. The chick starter will contain vitamins and minerals, protein, and fiber, all important ingredients for optimal chick growth.

Read the labels on the bags of feed to see what grains and ingredients are in the chick starter and what the percentage of protein is. Chicks need feed with a protein ration of between 18% and 20% for layers and between 22% and 24% for meat birds. The protein is essential for muscle development and supplies the chicks with amino acids that aid in the bird's growth.

Note the ingredients listed on the package and the order in which the ingredients are listed. By law, the most abundant ingredient must be listed first, with the rest of the ingredients in descending order.

The feed will contain soybean meal—dried, pulverized beans—which are a great source of plant-based protein. Another ingredient is canola meal—dehydrated, crushed seeds from the canola plant—which is high in protein. It contains cracked corn, which lacks protein but contains fiber and vitamins. Wheat middlings is a byproduct of wheat's milling process that is added to chick starter because they are a good source of fiber, protein, and complex carbohydrates to give the chicks an energy boost. Alfalfa is rich in nutrients, and calcium carbonate helps with bone and feather development. The feed will also include supplemental vitamins, minerals, and probiotics to satisfy the chicks' nutritional needs.

The label on the chick starter will provide additional information. It will tell you if the feed is medicated or natural, as well as if it is organic and non-GMO. Medicated chick starter contains an antibiotic and preventative medication to prevent coccidiosis, a parasite that attacks the birds' digestive tract. Natural chick starter does not contain this medicated formula, nor does it include antibiotics. To be able to carry the organic and/or non-GMO label, all the grains in the chick starter must be certified grown on organic and/or non-GMO farms.

Fresh Water

Chicks need clean, fresh water. Remember that the waterer likely sits underneath a heat lamp in temperatures that hover around 90°F. That is the perfect temperature for bacterial growth. When you refill your chicks' water, take a few moments to wash it out with soapy water and thoroughly rinse it before you fill it back up and return it to its spot in the brooder box. You will prevent the growth of slime, algae, and bacteria.

Many suburban homes have water softeners, so you may wonder if soft water is safe for your chicks. Water softeners use salt to remove the hard calcium and minerals in well water, as you know if you have to routinely lug a heavy bag of salt to your basement. Fortunately, the softening process leaves very little salt in the water that comes out of your tap. The FDA has even stated that soft water is safe for people under sodium restrictions to drink. So, in general, your soft water will be safe for your chicks to drink. There is one caveat to this.

Some chicken farmers like to keep supplemental electrolytes on hand to add to their chickens' water if the birds are stressed, overheated, or sick. Other people avoid using supplemental electrolytes, preferring to stick with plain water. If you need it, you can buy electrolyte packets at the feed store; it's not the same thing as

Gatorade. Anyway, electrolytes contain a lot of sodium. If you add that to soft water, you could be giving your chicks too much sodium, so keep that in mind.

A last point to consider about water is that chickens can be kind of picky about changes to their water. When my kids were involved in 4-H and statewide poultry shows, we always brought a jug or two of water from home when we went to a show. The birds sometimes refused to drink water that didn't taste familiar to them. If you start your chicks out by giving them soft water from your kitchen tap, then you switch to hard water from your outside garden spigot, your flock might not like the change.

Supplemental Treats and Grit

Young chicks have very specific nutritional needs that are best met by feeding them chick starter. It might be tempting to offer your chicks a tasty treat or two, but I would recommend that you avoid doing this until the chicks are at least three months old. Craig Hansen, a suburban chicken farmer from Iowa, agrees: "There are many quality starter feeds available," he explains. "They don't really need more than that to remain healthy and productive."

Photo Courtesy of Casey McMahon

Chicks have tiny stomachs. When you feed them a treat—even one as small as a blueberry—you are filling up their tummies so they are not hungry for the nutritional food they need. It only takes a few treats to throw off the balance of their diets. For now, let your chicks eat their baby food. There will be plenty of time for kitchen scraps when they are older.

We will talk more about grit in Chapter 5, but at this point in your chicks' lives, they don't need grit. Their chick starter is milled to such a fine texture that they are able to easily digest it. When they are older and living out in their coop, they will probably need grit to help them digest the food they eat, but not now. The only time that baby chicks need grit is if they are not fed a diet of chick starter.

Keeping Your Chicks Warm

Very young chicks need to be kept warm. The temperature in their brooder box should be between 90 and 95°F for the first week or so. After that, you can begin reducing the temperature by about five to eight degrees per week until it is time for the chicks to move outside to their coop.

Heat lamps can put off a lot of heat. Be sure to put a thermometer in the brooder box periodically to check the temperature. I use one of those outdoor thermometer things, the kind you hang outside your window. I don't leave the thermometer in the brooder box all the time. I put it in every so often to check the temp. The space under and around the heat lamps should be the appropriate temperature, but beyond that, it can be cooler. This will allow the chicks to step away from the heat if they get too hot.

Take extreme care with heat lamps. Make sure the lamp is secure so the chicks don't knock it over. It could cause a fire if the lamp's bulb comes into contact with the bedding material in the brooder box.

Provide Proper Ventilation

The chicks need to be kept warm, but they also need good ventilation. If the chicks become overheated, it could be deadly for them. They need to be able to cool off ... but not too cool.

Proper ventilation also protects the chicks from the effects of ammonia buildup in their brooder box. Chickens, like all poultry birds, have a waste product called excreta, which is a mixture of urine and feces. It is high in ammonia, which will soak into the bedding material. If the

bedding is not changed on a regular basis, the smell will become over-whelming and the gasses that it gives up can burn the chicks' eyes if there is not proper ventilation. Make sure the sides of the brooder box are not so high that airflow can't circulate down through it.

Monitor Your Chicks' Health and Growth

During the initial weeks of chick rearing, when your birds are still in their brooder box, you should monitor them to ensure that they are growing and thriving. This includes taking preventative health measures, watching for injuries or signs of illness, tracking their growth, and gauging when to move them to their outdoor coop.

Antibiotics? Yes or No?

In the past, both large commercial chicken farmers and small-scale backyard chicken farmers routinely gave newly hatched chicks a dose of antibiotics. Today, this is a more controversial topic. This issue lies in the fact that antibiotics were given as a preventative, not to treat an existing illness. The overuse of antibiotics in chickens, as in humans, leads to antibiotic resistance. So, what does this mean for suburban chicken farmers?

As of 2017, the United States Food and Drug Administration has stated that backyard chicken farmers can no longer purchase water-soluble antibiotics to give to their birds unless they have a prescription from a veterinarian. This is part of the FDA's Veterinary Feed Directive and is meant to decrease the amount of antibiotics that leach into the groundwater.

Although large factory farms with hundreds of chickens still give their birds routine, preventative doses of antibiotics, this practice is now frowned upon for backyard and suburban chicken farmers. It simply is not necessary because the flock is so small and relatively isolated. On a commercial level, antibiotics are necessary to keep diseases from ravaging an enormous flock.

Some chick starters contain non-water-soluble antibiotics, along with other medications. If you are concerned that your chicks need a preventative, this is your best option. Otherwise, most experts suggest that you only use antibiotics when necessary to treat an infection or illness.

Check for Injuries or Illness

Get in the habit of giving your new chicks a quick exam every day. Thoroughly wash your hands, then pick up each chick one at a time. I know from experience that it is hard to keep track of which chicks you have already examined. They run around and mix themselves up, and it is almost impossible to tell them apart. But do the best you can.

Look over each chick to make sure there are no open cuts or sores. Carefully spread out its wings, check its feet, and look at its eyes and nostrils. The chick's eyes should be bright and clear, and there should

Photo Courtesy of Tiffany Lawyer

be no drainage from the bird's eyes or nostrils.

Pay particular attention to the chick's vent or rear end. The vent is the opening through which the excreta comes out, and once the bird is a full-grown hen, it's where the egg will come out. Baby chicks are prone to a condition called "pasty butt." The excreta sticks to the fluffy down surrounding the vent and builds up. You will notice it right away. It looks like the chick has a dirty bum. If the dried excreta completely blocks the vent, it can be deadly for the chick. Pasty butt is usually caused by stress over temperatures in the brooder box; either the chick is too chilled or overheated. Only rarely is pasty butt a sign of infection or illness.

If you notice that one or more of your chicks has pasty butt, don't panic. You just need to clean the dried excreta off. Place the chick in a small bowl of warm water or gently hold its bum under the tap and run warm water over it. Allow enough time for the dried excreta to soften enough that it falls off on its own. Avoid pulling the poo off the chick's bottom. If it is stuck on tight, you may risk pulling off a chunk of the chick's skin in the process, opening up the bird to infection. Work quickly as you bathe the chick to make sure it stays warm enough, and try to towel dry it as much as you can before you return it to the brooder box. When all the chicks have clean bottoms, check the temperature in the brooder box and adjust accordingly.

Tracking Growth

As a new chick owner, you naturally want to make sure that the growth of your chicks is on target. If you ask any veteran suburban chicken farmer, they will tell you that the best way to gauge the growth of your chicks is to compare them to each other. If one is not thriving, it will be smaller and bonier than the rest.

But if you are a hard-data person, you may not be satisfied using the observation and comparison method. It is possible to track your chicks' progress by weighing them, but this is not an exact science. First, you need a small scale that measures in grams. After all, baby chicks weigh about as much as a dust bunny!

The rate of growth, the weight of chicks, and their overall size vary greatly from breed to breed. Some breeds grow more quickly than others. Some grow larger than others. The growth rate differs by gender too. There are a number of variables involved. But here is a ballpark.

The rate of growth should be in the range of 1.65 and 1.8 times every two weeks. For example, if a day-old chick weighs 35 grams, it should weigh between 57.75 and 63 grams at two weeks old and between 95 and 113 grams at four weeks old, and so on. This is just a general rule of thumb. If your chick is not hitting the benchmark, don't panic. As long as it is healthy, active, and thriving, it will be fine.

I have never weighed my baby chicks. I've opted to use the observation method. When you see your chicks every day, it is sometimes easy to forget how much they've changed and grown. I tend to take a lot of

DID YOU KNOW?

The Function of Feathers

A chicken's feathers do more than protect your bird's skin; feathers are critical for providing insulation, regulating your chicken's body temperature, and helping these birds fly. Baby chickens are born with a layer of fuff, but this baby fluff will be replaced with feathers around 5 to 6 weeks old. Then, these baby feathers will protect your chickens for their first 18 months, after which a cycle of feather loss and regrowth will begin. This process, called molting, can last from 8 to 12 weeks and occurs annually for most chickens.

pictures so I can look back on my camera roll to see how small the chicks were when they first arrived versus now. That's how I make sure they are all growing as they should be.

When Are My Chicks Ready to Move to Their Coop?

At some point, the chicks become too big for the brooder box, and they might even start flying out of the box. The brooder box setup gets smelly, despite how much you clean the box and replace the bedding material. After weeks of living indoors, it is time for them to move to their coop. Unfortunately, there are no hard and fast rules about the best time to move chicks from the indoor brooder box to their outdoor coop, but there are some things to consider, such as the feathering on your chicks, the outdoor temperatures, and factors in the coop.

Feathering

Newly hatched chicks are covered in a fine, fluffy down, but they quickly grow their feathers. It varies by breed, but most chicks are fully feathered by the time they are five or six weeks old. At that time, they are better able to regulate their body temperature.

Weaning Chicks from the Heat Lamp

We mentioned this in the previous chapter, but it is important to revisit it at this point. When you first bring your new chicks home to their brooder box, set up a heat lamp to keep the temperature between 90 and 95°F. Every five to seven days, you should adjust the heat lamp to decrease the brooder box temperature. If you decrease the temperature by five degrees each week, by the sixth week, the chicks have become acclimated to about 65 degrees.

Beginning in the third or fourth week, you should also shut off the heat lamp periodically so the chicks learn to deal with darkness. If the weather is particularly mild in your area, you could even shut the heat lamp off for hours at a time to help wean the chicks from the heat lamp.

What's the Weather Like?

Photo Courtesy of Kelly Kneeland

No matter the time of year, consider the weather and the forecast before you send your chicks outdoors. If the nighttime temperatures are well above freezing and daytime temperatures are in the 60s, the chicks will be fine. While it is important for young chicks to be kept warm, older chicks don't need constant warmth.

Make sure there are no storms predicted for your area when you relocate your chicks. They will have to get used to rain and thunder at some point, but let them get used to the big wide world first.

Coop Considerations

Inspect your coop before you take your chicks out. The chicks will still be smaller than full-grown hens, so make sure that they can't squeeze through the fencing or under a gate. You may need to reinforce some parts of the coop to prevent escapes or wait a bit longer for the chicks to get bigger.

FUN FACT
Chicken Math

According to a study out of Italy, baby chickens are capable of doing basic arithmetic. In this study, baby chicks were shown two groupings of objects. Then, the researchers obscured the objects with screens and moved one object from one group to another. The chicks were then able to locate the group containing more objects based on their observation of an object being moved.

Introducing Your Older Hens

Supervise your older hens around your chicks, at least initially. Hens have a complex social structure that includes a leadership hierarchy—yes, a pecking order. They don't take kindly to surprise interlopers. You probably won't find a mother hen that wants to care for the chicks. Instead, you will notice that the older hens bully the younger ones as a way to show them where they stand in the pecking order.

This is natural. You really can't do anything about it except rescue a chick if it is getting physically harmed. Most of the time, it doesn't dissolve into violence, but from time to time, a group of hens will target and pick on another chicken. They will peck at the victim's comb, waddles, or vent. If they draw blood, remove the victim until the wound heals. Remember that chickens are attracted to the color red. When they see blood, they will peck even more. Again, don't worry too much if your older hens and your chicks don't seem to get along. They will soon learn to live together in harmony.

Take the Chicks for a Test Run

Before you commit to moving your chicks to the outdoor coop, consider trying a test run. Take the

Photo Courtesy of Tiffany Lawyer

chicks outside on a sunny afternoon and let them run around in the grass. They will probably be a little hesitant at first, but in no time at all, they will be pecking at the ground, pulling at the grass, and soaking up the sunshine. Let them see the coop and your older hens. After a few hours, you can take them back to their brooder box. You'll be able to gauge from their behavior if they are ready to truly become backyard chickens.

Chapter 3 Summary

It is an exciting time when you first bring your new chicks home. Your dream of having your own suburban flock of chickens is becoming a reality. But before you can start gathering eggs, you need to raise your young chicks in a brooder box until they are big enough to live in an outdoor coop. Baby chicks need clean, fresh water, warmth, and nutritious food that is formulated for growing chicks. You will need to monitor their health and growth as well. You will be amazed at how quickly those fluffy little cotton balls turn into gawky teenage chicks as they get their feathers.

The next step is moving the chicks to their permanent home. In the following chapter, we will discuss the housing requirements for suburban chickens and the different types of coops that are commonly used.

CHAPTER 4

Housing for Suburban Chickens

> **"**
>
> *You will need a well-built, sturdy coop that is predator-proof. Premade coops usually won't last more than a year or two. They look cute but function poorly. I would rather have an ugly but well-made coop that functions versus a cute premade coop. If a coop functions well, chicken keeping will be fun. If you struggle with a coop that doesn't function and doesn't meet your and your chickens' needs, it won't be as much fun to keep chickens.*
>
> AMY JOHNSON
>
> **"**

The increased popularity of backyard chickens has opened up more options for coops and henhouses, but suburban chicken farmers still have some unique challenges to consider. In this chapter, we will examine the different types of housing for small chicken flocks and hear from some veteran suburban chicken farmers to get their expert opinions on what to look for in a coop. In the words of one of these experienced chicken keepers, Amy Johnson of Minnesota, "If a coop functions well, chicken keeping will be fun. If you struggle with a coop that doesn't meet your needs and your chickens' needs, it won't be much fun to keep chickens."

In addition to covering the pros and cons of the different housing types, this chapter will discuss the features that every chicken coop must have, tips for keeping the coop clean and safe, and the various bedding materials you can use.

As a suburban chicken farmer, your first step, before you shop for chicken coops, is to check with your homeowners' association, neighborhood association, and local ordinances to make sure your henhouse complies with the rules and laws. Some homeowners' associations, for example, prohibit residents from building permanent sheds and outbuildings. Others have rules about the size of coops and restrictions on where they can be placed. You certainly don't want to go to the expense and effort of constructing a chicken coop only to be told you must remove it.

Types of Chicken Housing for Small Spaces

Armed with your list of do's and don'ts from your neighborhood association and local regulations, you can begin to narrow down the type of housing you want for your new flock. There are prefab coops, custom-built coops, converted sheds, and mobile chicken coops. Each type has pros and cons, as our panel of experts will explain.

Pre-made Chicken Coops

Walk through your local farm and feed store or do some online shopping, and you'll see really cool-looking chicken coops that are sold as do-it-yourself kits or prefab housing. Some of them are quite elaborate and fancy. They are meant to enhance the aesthetics of your suburban backyard and become a focal point of your landscaping. I'll admit it, when I

Pros

- They are ready-made housing, so you don't need to have construction experience or be handy with wood and nails.

- They are cute, stylish, and attractive.

- They enhance the appeal of your yard.

- The coop will be ready almost immediately.

- They are not permanent, fixed structures, so many homeowners' associations allow them.

Cons

- They are expensive.

- They may not be well-built.

- They may not be predator-proof.

- They may not be functional.

- Some premade coops are too small for humans to get inside to clean.

- They are not permanent, fixed structures, so they are not anchored down in severe weather.

see chicken housing like this, I have coop envy. Of course, appearances can be deceiving.

"Premade coops usually won't last more than a year or two," says Amy Johnson of Minnesota. "They look cute but function poorly." When shopping for a premade coop, keep this in mind and look for one that is made with quality materials and is solidly constructed. Also, make sure that the coop you are looking at is secure enough to keep predators out. Here are some other things to consider.

Custom-built Chicken Coops

If you are a do-it-yourselfer or have a construction background, you can consider building your own henhouse. You can even hire someone to build your coop for you. Depending upon the level of your carpentry

Photo Courtesy of Mary Eaton
Custom Built Chicken Coop

Photo Courtesy of Mary Eaton
Custom Built Chicken Coop

Pros

- ⊘ You can get exactly what you want and need when you design and build your own coop.

- ⊘ They can offer more protection from predators.

- ⊘ If you build it yourself, it can be done inexpensively.

- ⊘ It can be a family project.

- ⊘ It can last longer and be sturdier than other housing types.

- ⊘ You can adapt the coop to your climate—more insulation in northern areas or more ventilation in warmer areas.

Cons

- ⊗ They can be costly if you have to buy all the materials and hire someone to do the construction.

- ⊗ They may not be as elaborate and attractive.

skills, your custom-built coop may not be as fancy and posh as a prefab one, but you will have the flexibility to create the type of housing that suits your needs and your situation. You even have the ability to customize some of the features to address problem areas.

For example, Elizabeth Sorby of Washington explains, "I designed my coop to have a wide door that could be kept open so the whole coop can be swept out into a large bin. The floor is lined with linoleum for ease of sweeping and cleaning. I've been so glad over the years that I took that extra step."

When we built our chicken coop years ago, we used dunnage—wood used as packing material—from my husband's workplace. The dunnage was headed for the dumpster, so we got it for free. He was even able to secure two floor-model double-pane windows that were also being discarded. We had recently replaced our front door, and we used it as the door of the coop. We had to break down and buy nails and shingles for the roof, but all said, it was a fairly cheap project that has stood the test of time.

Finding salvaged building materials is not too difficult. This is a route that many suburban chicken farmers take when preparing their chicken housing. William Quigley of Washington told me, "I built my last coop entirely from scavenged materials. Check websites and 'buy nothing' groups."

Converted Sheds, Outbuildings, Doghouses

Pros

- ⊘ If you already have a shed or outbuilding, you will save a lot of time and money.

- ⊘ You won't have to get special permission from a homeowners' association for existing buildings.

- ⊘ Aside from the doghouse option, sheds or outbuildings are tall enough for you to enter and clean.

Cons

- ⊗ If you are using a converted doghouse or kennel, it likely won't be tall enough for you to enter and clean it.

- ⊗ It may not be as aesthetically pleasing as a cute, small coop.

- ⊗ It will need some modification.

- ⊗ It may not have adequate ventilation or insulation.

Chickens aren't too particular about their accommodations. They will be happy as long as they have a secure place to go for the night that will keep them from freezing in the winter months. They don't require anything too fancy. If you have an unused garden shed in your backyard,

you can easily convert that into a henhouse. You can do the same thing with a small outbuilding or a large, roomy, kennel-like doghouse. You will only need to make a few small improvements to make these suitable for a small, suburban chicken flock, like adding roosts and nesting boxes. You will also have to make sure the shed is secure enough to keep out predators, as well as rats and mice.

Mobile Chicken Coops or Chicken Tractors

We have a chicken tractor in addition to our permanent chicken coop. We didn't really need it, but it was one of my husband's pandemic projects. Like all mobile coops, the chicken tractor is designed to be movable. The idea is that we are able to move the unit to a fresh plot of grass when the chickens trample, eat, and soil the current one. Most of these mobile coops, ours included, combine a house-like area that can be closed up at night with a fenced-in chicken run where the hens can hang out during the day.

Pros

- ⊘ Mobile coops can be moved to fresh grass.

- ⊘ They are small enough to use in a suburban setting.

- ⊘ They are easy to build and/or assemble.

Cons

- ⊗ They may not be predator-proof.

- ⊗ If you have a small yard, you may not have room to move it around.

- ⊗ They can be heavy and hard for one person to move.

- ⊗ They are not well-insulated for cold weather.

- ⊗ Some are not large enough for a person to enter and clean.

- ⊗ They can be costly.

Photo Courtesy of
Anne Greenwood

Our chicken tractor is kind of cool—my husband has to show it off to friends and family when they stop over—but there are some drawbacks to it. The biggest issue is that it is not predator-proof. Yes, the fence extends all the way to the ground, but our yard is not completely flat. There are dips here and there that create gaps under the fence. We learned the hard way that determined animals can wiggle through to get at our chickens.

Also, the house part of the chicken tractor was designed to have a wire mesh floor so bird droppings fall right through to the ground. The housing portion is secure and positioned a couple of feet off the ground, but a raccoon was able to get into the fenced part and reach up through the wire mesh to mutilate a few chickens. Our chickens now go into the regular coop at night.

Determining the Best Coop Size

"

Coops are either chicken-scale or human-scale. I am not a fan of human-scale coops for urban environments. The only advantage is that you can step inside for cleaning, but there are many disadvantages—for one, it's very difficult to make a large coop rat-proof. Even the smallest coop that you can step inside will have far more space than required by the number of hens your city most likely allows (eight here in Seattle), and all that wasted space takes away from precious urban gardening space or just backyard area.

WILLIAM QUIGLEY

"

A crowded coop can cause all sorts of problems for your chickens, so it is important for you to get the correct size coop for your suburban flock. Since the size is based on the number of chickens you have, you will need to work within the allowable number of birds permitted by your neighborhood association or community ordinances. For example, if your community only allows up to five chickens per household, there is no need for you to erect a chicken coop that can accommodate 20 birds. You will also have to take into consideration the rules for coop sizes. Some communities or subdivisions have restrictions on how large a chicken coop can be.

The standard rule of thumb for determining the size of a chicken coop is to allow three to four square feet per bird. With this in mind and knowing that you are only allowed to have a maximum of five birds, for example, you should plan for your chicken coop to be between 15 and 20 square feet. If you go any smaller, you risk your birds fighting. Hens can become mean and aggressive if they are crowded and don't have enough personal space.

Needs and Wants: What Every Coop Must Have—And Features that Are Nice to Have

> **"**
>
> *Think about how you want to collect eggs and clean out the coop. Do you want to be able to reach the eggs from outside of the coop/run, without having to walk through muck (especially in the winter)? Cleaning out the inside of the coop also needs to be done periodically. I designed my coop to have a wide door that could be opened and allow the whole coop to be swept out into a large bin. The floor is lined with linoleum for ease of sweeping. I've been so glad over the years that I took this extra step.*
>
> ELIZABETH SORBY
>
> **"**

Your suburban chicken coop must have a few features to meet the needs of your hens, but there are some other features you may want to have to make chicken farming more convenient for you and your family.

Needs

> "
>
> *Make sure they have plenty of access to food, water, and nesting boxes and they will come into the coop when they need to. I also recommend building a coop that's around a foot off the ground. This helps keep predators out, as well as rats. Rats are the biggest problem for the urban coop because they are ubiquitous, and keeping chickens can also mean feeding all the neighborhood rats, which can spread disease to your flock.*
>
> WILLIAM QUIGLEY
>
> "

Let's start with the needs. It is an absolute must that your chicken coop has a door that completely closes, roosts, nesting boxes, protection from predators, good ventilation, and sunlight.

A Good Door

This should actually say "doors." Many chicken coops, mine included, have two doors, a chicken door and a people door. Smaller coops and mobile chicken tractors may only have a chicken door. Either way, all doors should close securely and latch tightly to keep your hens safe. We used to have a hook-and-eye lock on the outside of our chicken door, but we have a raccoon problem from time to time. Raccoons have dexterous little hands. They are able to pull the hook out of the eye pin to get into the coop.

You can purchase and install an automatic chicken door on your coop. These are set to close after sunset when your chickens will have gone into the coop to roost. While it would be really convenient not to have to go outside every night after dusk to close up the coop—and we have admittedly forgotten a few times—I have not personally used an automatic chicken door. My worry is that a chicken might lag behind and not go into the coop with the others at dusk. If the door closes automatically, that hen will be stuck outside all night.

Roosts

Chickens have a strong roosting instinct. This trait was ingrained in chickens long before they were domesticated and is one of their survival tactics. Every evening, they seek out high perches or roosts to sleep on. The higher, the better. They do this to stay as far away from predators as possi-

Photo Courtesy of
Brian and Jen Jacob

ble. Make sure that your coop has roosts; a two-by-four in the rafters or an elevated bar across the coop will suffice. If your coop does not have roosts, your chickens will seek out their own roosts. You may find them in a tree branch or on top of a fence when the sun goes down instead of going into their coop because they have a strong urge to sleep on a high perch.

Nesting Boxes

You are starting your suburban chicken flock because you want fresh eggs, so you need to plan a place for your hens to lay their eggs. Your coop will need to have some nesting boxes.

Hens prefer a little privacy when laying their eggs. They will instinctively find a quiet, secluded spot to lay. It could be behind a bush, under your deck, in some tall grass, or, ideally, in a nesting box. Unless you want to make every day an Easter egg hunt, you need to provide nesting boxes in your coop.

Each nesting box should be about one foot wide by one foot deep with a height of one foot. This will give a hen adequate space to do her duty and still give her that secure, enclosed feeling she needs. Mount the nesting boxes off the ground, too, as hens feel safer when they are higher up.

You don't need to supply a nesting box for each hen. In fact, the standard ratio is to provide one nesting box for every three or four chickens. For a small suburban flock with only five or six hens, having two nesting boxes is more than adequate.

Fill the nesting boxes with straw or another type of soft, absorbent material to make the hen comfortable and cushion the

Photo Courtesy of
Lana Plashchynskaya

eggs so they don't break. Hens typically start laying between the ages of 18 and 22 weeks old. As they get closer to that age, make sure the nesting boxes are ready and periodically put the hens, one at a time, in the boxes. They will probably flap and jump right out, but they will remember the place when it comes time to lay their first eggs.

Protection from Predators

"Safety is the number one priority!" says Amy Johnson of Minnesota.

We have mentioned predator protection several times, and we will devote an entire chapter to it later in this book. It is worth including a brief overview at this time because security is a coop issue. The coop should not have any cracks or gaps in it. Even if a larger predator, like a coyote or fox, can't get in, a smaller one, like a snake or a rat, can. All the entry points—doors, windows, nesting box lids—must be able to close tightly and seal completely. If the coop has a dirt floor, consider lining it with wire mesh to keep burrowing animals from digging their way in.

The fenced-in portion also needs to be completely secure and free of gaps and openings. Chicken wire is rather flimsy; some predators can bite through it. You could double up the chicken wire or invest in

a heavier-grade fencing material. Pay particular attention to the area where the fence meets the ground, as this is a common place for animal invasions. Fill in all the gaps and inspect the area on a regular basis to check for signs that an animal has tried to dig its way under.

Proper Ventilation

As Elizabeth Sorby of Washington explains, "Chickens are built for the cold weather as long as they can get out of the wind." High heat and humidity are more detrimental to your flock, which is why it is vital that your coop be well-ventilated. Good airflow through the coop will reduce the moisture and humidity and lower the temperature. Proper ventilation helps remove dust from the air, as well as reduce airborne diseases. Ammonia buildup causes irritating fumes in the coop, but airflow will prevent excess fumes.

Natural Sunlight

Plenty of sunshine is not just good for your chickens' mood, but it is a factor in their egg production, as we will see later in this book. Hens need a certain number of daylight hours to continue laying eggs, which is why most chickens stop laying over the winter. Plan your coop to have windows or skylights to allow for natural sunlight to come in. Additionally, make sure your hens get outside, even in the colder months, to soak up the sun.

Wants

> 66
>
> *Think about function. Make the run tall enough to stand in. No one wants to hunch over each time they clean the run. Do you want electricity? In Minnesota it gets dark early, and I enjoy a light in my coop for when I do chores after dark. Electricity also provides heat, heated water dishes, supplemental lighting to encourage egg laying over the winter, and an automatic coop door. Humidity is also a concern; ventilation is a must.*
>
> AMY JOHNSON
>
> 99

Now that we have covered the features that must be in your suburban chicken coop let's move on to discuss the features that would be convenient or handy but aren't absolute musts. These include people doors, exterior egg collection lids, exterior feed tubes, and electricity.

A People Door

Many small suburban chicken coops, especially mobile coops and premade coops, are designed for only chickens to enter. They simply are not large enough for people to go into them. This makes cleaning the inside of the coop a tremendous challenge.

In a suburban setting, a coop that is tall enough for an adult human to stand inside can be too large for the space. William Quigly of Washington prefers smaller coops. He explains, "I'm not a fan of human-scale coops for urban environments. The only advantage is that you can step inside for cleaning." Having chicken housing that is large and roomy enough for one or two people to go inside is not a necessity, but it is a convenient luxury that makes suburban chicken farming easier and more enjoyable.

Exterior Egg Collection Lid

You can find premade coops, or design your own, that offer a way to collect eggs from the nesting boxes without going inside the coop. The nesting boxes are built to jut out from the side with a lid like a chest. To collect the eggs, you don't need to go inside the coop at all; you just need to lift the lid to have easy access to the nesting boxes.

Exterior Feed Tubes

"You may also want to consider how you want to feed your chickens," says Elizabeth Sorby of Washington. "I've seen some great coops that have tubes for adding chicken feed from the outside, so you don't have to enter the run or the coop. That would

be great in the winter when things get pretty muddy." Some premade chicken coops include these exterior feed tubes, but some clever suburban chicken farmers have constructed their own exterior feed tubes using PVC pipes. It is convenient to be able to pour a scoop of feed into the pipe without entering the coop, but there are some drawbacks. There needs to be a closure on the tube so that mice or rats can't scurry in looking for food. It also doesn't give you a chance to gauge how much food your chickens are eating and discard old, spoiled, uneaten food.

HELPFUL TIP

Odor Control

According to a study out of Italy, baby chickens are capable of doing basic arithmetic. In this study, baby chicks were shown two groupings of objects. Then, the researchers obscured the objects with screens and moved one object from one group to another. The chicks were then able to locate the group containing more objects based on their observation of an object being moved.

Electricity

Humans have kept domesticated chickens for centuries without the need for electricity, but it is a nice convenience. "In Minnesota, it gets dark early," explains Amy Johnson. "I enjoy having a light in my coop when I do chores after dark." Electric lights make it easier for you to see what you are doing in the early morning or in the evening. You could use a heated waterer to prevent your chicken water from freezing if you have electricity in your chicken coop. You could even install electric lights that replicate natural sunlight so your hens have enough light to keep them laying eggs all year long. It would even be possible to power an automatic coop door, an alarm system, and a coop camera.

You cannot run electricity to a mobile chicken coop or chicken tractor; it would only be possible to add it to a permanent structure. There is an initial cost to hire an electrician to wire your coop and an increase to your monthly electric bill. You will also need to take precautions to prevent an electrical fire, including covering all wires so the hens don't peck at them.

*Photo Courtesy of
Brian and Jen Jacob*

The Chicken Run

The chicken coop has a chicken door to allow the hens out—but to where? That depends on how you feel about free ranging versus using a chicken run, a topic we will delve into more in Chapter 8. Your coop may open into your yard or garden, where your chickens can spend their days exploring, or it may open into a chicken run, a fenced space where your hens can spend time outdoors protected from predators.

If you have a chicken run attached to your coop, be sure the fencing material is sturdy, durable, galvanized wire. The fence should be supported by weather-treated wood or metal posts that have been securely erected. There should be no holes or openings in the fencing. If at all possible, the fencing material should also extend across the top of the chicken run to prevent hawks and owls from picking off your hens.

Bedding Material for the Chicken Coop

> **"**
>
> *I use a highly compostable mulch as 'litter' in my run, and when I feed the chickens treats, I toss the treats into the middle of the run. They either eat them or the treats break down in the run—it functions like a small compost pile. I shovel out the run once a year in winter and dump the nitrogen-rich compost in the flower beds, where it has a few months to break down before spring.*
>
> WILLIAM QUIGLEY
>
> **"**

Because chickens roost on their perches when they sleep, it might be easy to dismiss the importance of bedding material in your chicken coop. The bedding is not necessarily meant to serve as a cozy bed for your chickens. Instead, it is litter that collects and absorbs chicken droppings. You have several options when it comes to bedding material, from straw to wood shavings to shredded paper. Let's look at these choices.

Straw or Hay

Straw and hay have been traditionally used as bedding material for chickens, and it remains a go-to material for coops. Bales of straw are relatively inexpensive, and if stored in a dry place, the straw would keep for quite a while. You don't need a lot of straw to fill a small coop, making it a viable option for suburban chicken farmers. Hay and straw are also compostable, so you can put the soiled bedding into your compost pile or directly onto your garden beds.

Wood Shavings

You can buy small quantities of cedar or pine shavings at pet stores or a large bag of them at a farm and feed store. Both pine and cedar shavings absorb droppings and dry quickly. As an added bonus, the wood shavings give off a clean, pleasant smell. Like straw, wood shavings can be used in your garden. Some chicken farmers believe that cedar shavings can cause respiratory distress in chickens, but there is no hard evidence to back this up.

Shredded Paper

One way to recycle old newspapers or scrap paper from your office is to shred it and use it as bedding material in your coop. The nice thing about this material is that it is free. Newspaper is more absorbent than shiny computer paper, but both will do the trick and will quickly decompose in your compost heap. If you are planning to use shredded paper as a bedding material, you trade the free cost for added labor. You will have to shred the paper yourself, either by tearing or cutting it by hand or by using a paper shredder.

Photo Courtesy of
John and Kim Anderson

Lawn Waste

You can use all-natural lawn waste, such as grass clippings and dried leaves, as bedding material. These items are free and offer a way to recycle your yard waste. These materials easily break down, making them easy to compost, but the material can start decomposing when it is still in the coop. Grass clippings are not very absorbent, so moisture lingers in the coop. It is also important that the grass clippings come from a lawn that has not been treated with chemical fertilizers or pesticides.

Sand

Sand is, initially, the most expensive option for coop bedding, but it lasts a lot longer. Think of it as kitty litter. You will need to regularly scoop out the clumps of droppings from the bedding material. If you are diligent about routine cleanings, you may only need to replace the sand a few times a year. Unlike other bedding materials, however, sand cannot be placed in a compost and then used in your garden. Also, the wet sand will freeze, making it difficult to scoop in the wintertime.

Chapter 4 Summary

Throughout this chapter, you have learned about the various types of housing options for your chickens, from pre-made designer coops to mobile chicken tractors and converted sheds. More important than the style of the coop, however, are the features. All coops must have certain features, like nesting boxes, secure doors, and roosts, but some optional features, like electricity, exterior access to nesting boxes, and people-sized doors, make suburban chicken farming more convenient. Lastly, we discussed the different materials that can be used as bedding in your chicken coop, including straw, hay, sand, and wood shavings.

The information outlined in this chapter will help you make informed decisions about the size, style, and type of coop you will use to house your suburban chicken flock, an important initial step to adding a backyard flock to your home.

CHAPTER 5

Feeding Your Flock

I n a previous chapter, we discussed feeding chicks and the nutritional requirements of young chickens, but once they reach about 18 to 20 weeks of age, it is time to take them off their baby food. Stroll through the feed aisle at your local farm store, and you'll be overwhelmed by the selection of chicken feed. In addition to a variety of brands, you'll find bags of feed labeled "grower," "layer," "finisher," "pellets," "crumbles," "scratch," and more. It can be hard to know what is best for your suburban chicken flock.

We will sort it all out in this chapter by explaining each type of chicken feed and the reasons why you would use each one. We will also cover the nutritional requirements of adult chickens and egg-laying hens. This chapter will also discuss the other things your chickens might eat, like insects, grass, kitchen scraps, yard clippings, and treats. Lastly, we'll talk about the feasibility of free-ranging your suburban chickens, as well as the pros and cons of this popular practice.

The Digestive System of a Chicken

Before we jump into the nutritional requirements of chickens, let's go over the unique yet highly effective digestive system of chickens. You know that old-timey saying, "That's as scarce as hens' teeth"? My grandma used to say that all the time, and as a kid, I wondered just how scarce chicken teeth really were. The answer is extremely rare—actually, nonexistent. A quick look inside your chicken's beak will confirm this. Chickens do not have teeth. So how do they chew their food?

The commercial chicken food you feed your hens is already quite small, but your suburban flock will also find tasty treats to eat around

ARE CHICKS BORN WITH A TOOTH?

If you take a close look at the beak of a newly hatched chick, you'll notice a tiny nodule protruding from its beak. This is an amazing adaptation affectionately known as an egg tooth.

An egg tooth is a small, sharp, spike-like structure on the top of a baby chick's beak. It is made of keratin, the same material that makes up our fingernails and hair.

As small and unassuming as it appears, the egg tooth is a unique adaptation designed to help the chick know when to bust out of its shell and how to do it. The egg tooth develops right before the chick is ready to hatch. This timing is perfect. If the egg tooth formed earlier, a chick might be able to make its escape prematurely.

The egg tooth is not just a timer; it is a tool. When a chick is ready to hatch, it will use its egg tooth to break a small hole from inside. Once the shell has been breached, the chick will use its beak to widen the hole until it can escape. The egg tooth is designed to be strong enough to puncture the shell but not so sharp that it will damage the young chick's skin.

Once the chick's egg tooth has done its sole job, it is no longer needed. Within 24 to 48 hours after the chick emerges from its shell, the chick will shed its egg tooth.

your yard, like grasshoppers, worms, grubs, crickets, seeds, foliage, and maybe even your prized tomatoes. I have witnessed my hens eating a frog on more than one occasion, and my husband claims one of them ate a small snake! Without teeth to chew these items, how do they digest them? With their gizzards!

At the base of the chicken's neck is its crop. Food moves from the bird's mouth, down the esophagus, and into the crop, a sack-like storage hold. The food is stored in the crop and, bit by bit, released to the stomach, or gizzard, of the chicken. The chicken's stomach is not like our stomachs. Human stomachs don't have to work that hard to break down food because we thoroughly chew our food before we swallow it. Chickens' gizzards break down food using a combination of digestive enzymes and physical grinding with grit.

The grit—small pebbles or sand—works with the chicken's stomach muscles to grind up grain into digestible particles within the gizzard. After the food is ground up by the gizzard, it can continue through the chicken's digestive tract.

Chickens intuitively know that they need grit in their gizzards to aid in digestion. When they peck the ground, they pick up tiny bits of dirt, sand, and stones. I have noticed our chickens hang out in our gravel driveway from time to time, pecking at the ground. Commercial chicken feed includes grit, which should suffice.

The Nutritional Requirements of Chickens

The nutrients and minerals that your chickens need vary depending on the age of the birds and whether or not they are laying. Young chicks, like all babies, need more nutrients, protein, and calories to help their tiny bodies grow and give the birds the building blocks they need for strong bones, healthy feathers, and active muscles. Once hens begin laying eggs, they need a tremendous calcium boost so they can form strong shells around their eggs. Chickens raised for their meat need to consume compounds that are vital for muscle growth and development.

According to agricultural biologists, chickens require a combination of 38 different nutrients for optimal growth and health, but since many of them are trace minerals, we will focus only on a few of the key requirements.

Protein

As my daughter's blue ribbon–winning fourth-grade science fair project confirmed, chickens are omnivores, meaning they eat meat in

addition to plants and grains. They require protein in their diets because the amino acids in protein are vital for the birds to grow their feathers, beaks, cartilage, skin, and more. This is especially important for young, growing chicks and for egg-laying hens. There are several different types of amino acids; therefore, commercial chicken feed mixes contain protein from a variety of sources, so the feed includes all the necessary amino acids.

Animal feed companies are required by law to provide the nutritional breakdown of their products on the labels. Read the label to find out the percentage of crude protein in the feed. The protein in chicken feed is usually a combination of plant proteins and animal proteins. Corn gluten meal, soybean meal, and canola meal are grains that are high in protein. Animal meat and bone meal make up the remainder of the protein in chicken feed.

Carbohydrates

Chicken feed is composed mainly of carbohydrates. The starches and sugars in carbohydrates provide the energy needed to fuel the chicken. Commercial chicken feed has large amounts of corn, barley, wheat, and other grains that are loaded with carbohydrates and easy for chickens to digest. Not all grains are created equal. Some of them contain too much crude fiber or cellulose, which chickens cannot digest well.

On the feed bag label, you will not see a percentage of carbohydrates like you see the protein percentage. But if you look below the nutritional breakdown, you will see a list of ingredients. Grains like corn and soybeans are listed first, meaning those are the most abundant items in the feed.

Fats

Fats not only make the feed tastier for your chickens, but they are a required element in the birds' diets. Chickens must have fats in their feed so their bodies can absorb fat-soluble vitamins. Chickens require linoleic acid, a fatty acid, so commercial poultry feed includes sources of fats that contain this acid. Chicken feed will include animal fat and yellow grease, but you won't find vegetable oils used in commercial feeds. It is simply too costly to use without drastically increasing the price of a bag of feed.

Minerals and Vitamins

Minerals in foods perform a number of different functions, including regulating metabolism, helping the blood carry oxygen, and aiding in the production of enzymes. Chickens, like all animals, need a balanced blend of minerals to keep their bodies functioning properly. For poultry, phosphorus is essential for bone development and the creation of cell membranes. On feed bag labels, you will see a phosphorus percentage listed. Likewise, the percentages of zinc and magnesium are also listed.

Chickens also need a diet that includes copper, iodine, chlorine, sodium, selenium, and manganese.

Grains do not contain a lot of minerals. If the mineral requirements are not met by the animal proteins and fats that are included in commercial chicken feed, the manufacturer will add a premix of minerals to the recipe to ensure that the nutritional requirements are satisfied.

Vitamin deficiencies can cause disease or health conditions. Your suburban chickens will need water-soluble vitamins, such as vitamin C and the various B vitamins—B12, niacin, riboflavin, thiamin, and so on. Chickens also need vitamins A, D, E, and K for the growth of skin and organ tissue, bone development, blood clotting, and eggshell production. Alfalfa meal is a good source of vitamins; therefore, it is included in chicken feed.

Calcium

Calcium is necessary for bone formation in all animals, including humans and chickens. Calcium also helps muscles function well and plays a vital role in allowing blood to clot. For hens, however, calcium is also necessary for eggshell production. Crushed oyster shells or pulverized limestone are commonly added to chicken feed to fulfill the calcium requirement.

We will hit on this a bit more when we discuss food supplements later in this chapter, but if you notice that your hens are laying thin-shelled eggs that easily break or if you find shell-less eggs—yes, that happens!—it is a sign that your chickens are not getting enough calcium in their diet. They need a calcium supplement, such as crushed oyster shells.

Water

You may not think of water in regard to nutritional requirements for chickens, but clean, fresh water is one of the most important parts of your flock's diet. Your chickens should have constant access to clean water. In fact, if your hens run out of water even for a few hours, it may interrupt their egg production for that day.

Chickens rely on water to soften the grain they eat and to help carry food particles through their digestive system. Since chickens don't have sweat glands, they struggle to cool themselves when temperatures rise. They need to drink more water to prevent them from overheating.

You can purchase automatic chicken waterers to provide continuous access to water. Or you can use regular waterers and commit to filling them at least twice a day. This is what I do with my chickens, and I can honestly say that it is a quick and easy task. I have a couple of extra waterers, so I don't have to walk out to get the container, bring it into the house to refill it, then take it back out. I just make one trip out to drop off the fresh water and bring the empty one inside for the next watering time. This method also works well in the winter when the waterers freeze. I put the frozen one in the laundry tub to thaw out while the chickens enjoy the fresh water.

As mentioned earlier, the nutritional requirements change throughout your chickens' life span. The chart below shows you the requirements by age for protein, calcium, and phosphorus, as well as the type of chicken feed based on the age of the birds.

Types of Chicken Feed

When shopping for chicken feed, you will find a number of options. Some of the options are geared toward the age of the bird, while others are simply different textures. Let's look at the different types of chicken feed.

Bird Type	Age	Food Type	Protein %	Calcium %	Phosphorus %
Egg Layers					
Pullets	0-6 weeks	Chick Starter	20-22%	.85-1.0%	.40-.45%
Pullets	6-14 weeks	Chick Grower	16-18%	.80-.95%	.35-.42%
Pullets	14-20 weeks	Finisher	14-16%	.75-.92%	.30-.38%
Layers	20+ weeks	Layer	15-19%	3.60-4.20%	.32-.40%
Breeding	mature	Layer	14-18%	3.40-4.00%	.32-.40%
Meat/Dual Purpose					
Pullets/Cockerels	0-4 weeks	Chick Starter	20-23%	.901.0%	.42-.45%
Pullets/Cockerels	6-12 weeks	Chick Grower	19-20%	.86-.92%	.38-.43%
Pullets/Cockerels	12-22 weeks	Finisher	15-18%	.78-.88%	.32-.40%
Layers	22+ weeks	Layer	14-16%	3.00-3.40%	.34-.41%
Breeding	mature	Layer	14-16%	2.75-3.50%	.30-.40%

Chick Starter

We touched on chick starter earlier in this book. This is baby food that has been specially formulated to meet the nutritional requirements of young, growing chicks in much the same way that puppy chow has been created to satisfy the unique nutritional demands of growing pups. Chick starter is higher in protein than adult chicken feed formulas and contains much less calcium because the baby chicks are not yet laying eggs. It is recommended that you feed your young birds chick starter at least until they reach the age of six weeks.

Chick Grower

Once your little flock reaches about six weeks of age, the birds are no longer babies. Their nutritional needs are evolving, and you can switch them to a chick-grower formula of chicken feed. The major difference between chick grower, sometimes called chick developer, is that it has a lower percentage of protein. While chick starter typically has a protein percentage of 20%, chick grower's protein level is about 15%. Not every suburban chicken raiser uses chick grower. Some people prefer to keep their chicks on chick starter until they reach adulthood.

Chick Finisher

Chick finisher is also called broiler finisher or meat-bird feed and is primarily used for chickens raised for their meat. When the chicks reach about six weeks of age, they are moved off chick starter and onto chick finisher, a feed that is formulated for ideal muscle growth to beef up the birds before slaughter. The feed is high in protein and carbohydrates, in addition to vitamins and minerals. Eating a diet of chick finisher gives the birds a boost of energy so that they are more active, thus building up their muscles. Most suburban chicken farmers, however, are keeping chickens for their egg output, not for slaughter. Therefore, you may never need to feed your flock chick finisher.

Layer

When pullets reach the age when they start laying eggs, which is between 18 and 20 weeks, depending upon their breed, they require much more calcium in their diets. Eggshells are composed of calcium carbonate that is formed in the hens' shell glands, but they need all the calcium they can get to ensure that they can produce quality shells.

Styles of Chicken Food

Crumbles

Chick starter is made in crumble form because the tiny particles are easier for small chicks to pick up and eat. Since this is the type of food that your birds will become accustomed to, you may want to continue to feed them layer crumbles when they are older. Like chick starter, layer crumbles are finely pulverized feed. It is important to note, however, that  crumbles and pellets are identical, nutritionally speaking. In fact, chicken feed manufacturers make all their food in pellet form. To make crumbles, the pellets are crushed.

Pellets

Chicken pellets resemble small tubes, but they are solid, not hollow. They look the same as rabbit feed. In the manufacturing facility, the feed mixture is pressed through molds, then baked hard. They easily break, so the pellets are usually smaller than a half-inch long. The pellets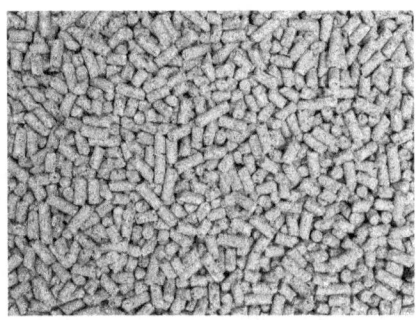

are less messy than the crumbles, and some suburban chicken farmers feel like there is less waste with pellets.

I have never personally had this issue, but I have heard that some chickens balk at transitioning from chick starter crumbles to larger pellets. Since crumbles and pellets contain the same ingredients and are identical in their nutritional value, it is really a matter of preference if you feed your flock crumbles or pellets.

Mash

Mash, like crumbles, is chicken feed in fine, small particles. Mash is composed of the same ingredients that you will find in crumbles and pellets, so the nutrient content is identical. The difference between mash and crumble comes down to how the feed is manufactured.

As noted above, the crumbles are chicken feed that was first made into pellet form and then crushed into the crumbles. The various ingredients have been blended and combined to form the pellets that form the crumbles. It is like the different ingredients that go into making a cake— flour, sugar, eggs, baking soda, etc. Imagine them being baked into a cake and then smashing the cake into crumbs. That's how pellets and crumbles are. With mash, however, the individual ingredients, like corn and alfalfa, are pulverized into tiny bits and stirred together in a fine, powdery mixture. Picture trail mix. It is a blend of different foods, like granola, raisins, and nuts, but we can still easily identify and pick out each one. Unlike the baked cake, the ingredients haven't merged together. That's the main difference between crumbles and mash.

Nutritionally, the crumbles and the mash are the same, but your hens will notice the difference. When they eat one piece of crumble, it is comprised of the same stuff as every other piece of crumble in the feed tray. With mash, however, a hen can pick out her favorite ingredients and skip over the ones she doesn't like. Knowing this, you can see that there are two problems that arise when feeding your suburban chickens mash. First, there is more waste, especially if your chickens are picky eaters. Second, your hens may not get all the nutrients they need if they are picking and choosing their favorite bits of food to eat.

Hot Lunch

One last word about chicken mash before we move on. The makeup and texture of chicken mash are ideal for making hot mash for your suburban chickens. On a frigid winter day, your hens will greatly appreciate a hot lunch. Just add some boiling water to the mash and stir it into a porridge-like consistency. Take it immediately out to your hens—it won't be too hot for them as it will start to cool as soon as it hits the cold air—and watch them devour it. Adding hot water to layer crumbles also works, but it doesn't create the same consistency as mash. Warm mash is a great way to help your birds endure the winter months and show them how much you love them.

Amy Johnson of Minnesota does this with her flock of suburban chickens. "In the winter, I bring them a warm bowl each morning," she says. "I use a heated dog water dish to keep it from freezing."

Scratch

Chickens like to scratch. The ground, that is. All chickens have an instinct that causes them to scratch the ground with their feet and peck at the dirt in search of tasty morsels to eat. In doing so, they also eat the dirt, sand, and pebbles they need for grit.

At the feed store, you will see sacks of chicken feed that are labeled as "scratch." This is a blend of corn, whole grains, legumes, and seeds that you can scatter on the ground for your chickens. Your hens will love it. It gives them an opportunity to satisfy their scratch urges and get some tasty treats. But scratch should be viewed as just that—a treat.

Scratch mixes are not formulated to meet all the nutritional needs of your chickens and should not be used as a main feed source for your flock. If they were to only eat chicken scratch, your birds would become deficient in certain areas, which could lead to health problems.

Like mash, the individual pieces in scratch feed maintain their originality. Your hens can pick out the pieces of corn, for example, and leave the rest. This is another reason why scratch should only be used as a supplemental feed.

How do I use scratch with my ladies? Every morning, when I open the coop to let them out, I take a scoop of scratch and scatter it on the ground in a different spot than the previous day. I rotate the scratching grounds. This encourages the hens to leave the coop, especially on colder days. Plus, it gets them out of my way and keeps them occupied while I fill their waterer, give them their layer pellets, collect eggs, and clean out the coop if necessary.

Like all treats, scratch should be used sparingly. Chickens love it and will gorge themselves on scratch, leaving no room for their healthier, more nutritious main course.

Supplementing Your Chicken Feed with Treats and Greens

Chickens love dried mealworms or grubs that can be purchased almost anywhere that chicken feed can be bought. It is incredibly useful to have grubs on hand if you need to collect the chickens back into the coop, as they'll follow you anywhere for a handful. They also love kitchen scraps of most kinds. I have both a compost bowl and a 'chicken bowl' in my kitchen for collecting scraps that I give them each evening.

ELIZABETH SORBY

As Craig Hanson of Iowa explains, "There are many quality chick starters and complete feeds available. Chickens don't really need more than that to remain healthy and productive." A rich and varied diet will keep your hens happy and thriving, which is why many suburban chicken farmers supplement chicken feed with greens. Their regular chicken feed, which has been formulated to satisfy their nutritional requirements, coupled with the food items they forage and occasional treats to supplement their diet, will be more than enough to do the trick.

Let's talk a bit about treats. Not all treats are created equal. Some are filled with empty calories and non-nutritious ingredients, while other treats can enhance and boost your birds' nutrient intake. In this section, we will discuss some healthy treat options you can try, as well as foods to avoid giving your hens and share some of the experiences from our panel of experts regarding supplemental treats for your suburban chickens.

"My chickens love dried mealworms or grubs that can be purchased almost anywhere that chicken feed is sold," explains Elizabeth Sorby of Washington. "It is incredibly useful if you need to round them up to go back to the coop. If you have these on hand, the chickens will follow you anywhere."

Fruits and Vegetables

> *My chickens eat regular feed from the farm store but also receive food scraps. I would recommend tossing your chickens the majority of your food scraps because it can be healthy for them and create great compost. Foods I have found my chickens love are watermelon, tomatoes, berries, warm cooked oats in the winter, and pumpkin.*
>
> KRISTEN HARRELL

Although the bulk of your chickens' diet consists of grains, they will welcome the opportunity to try some tasty fruits and vegetables. These are packed with vitamins and minerals in a different texture and form than the crumbles or pellets of the hens' regular food.

Ones to Try

- **Strawberries** – Chickens are attracted to the color red, so they will have no trouble spotting delicious, juicy strawberries. The seeds of the strawberry are tiny enough to be easily digested. Strawberries should be given to your chickens in moderation, as they are high in sugar. If you grow strawberries in your yard, you may have to devise ways to keep your hens from eating all the berries. Given half a chance, they will!

- **Blueberries** – Blueberries are a superfood that is high in nutrients and antioxidants. They are great for your suburban chickens as well. The size of the blueberries is ideal for your hens too.

- **Pumpkin** – Pumpkin is a natural dewormer that your hens will love. They can eat the skin, flesh, and seeds of the pumpkin. You can give your flock one whole pumpkin for hours—or days—of enjoyment. They will peck at it until they break open the skin and can get to the flesh and seeds inside. On more than one occasion, my chickens have destroyed the jack-o'-lanterns we carved and set on the porch steps for Halloween. But I avoid giving my chickens the carved pumpkins after Halloween is over. Why? Two reasons. First, the insides are usually borderline rotten by the time November 1 rolls around. Second, the candles that we light inside the jack-o'-lanterns turn the inside black and sooty, and there are always gobs of wax clumped on the bottom. None of that sounds appealing to feed my chickens.

- **Watermelon** – Watermelon and watermelon rinds are tasty treats for your chickens. They are even able to digest watermelon seeds. You can either cube up the watermelon for your hens or toss a few of the rinds, with a good amount of the watermelon still on them, to your flock. They will entertain themselves pecking at the fruit, but they probably won't eat the outer skin of the rinds. You'll need to pick those up later and discard them.

- **Bananas** – Bananas are another good source of vitamins and nutrients, especially vitamins A, C, and B6. Chickens, however, don't care for banana peels, nor can they peel their own bananas. But if you peel one for them, they will happily enjoy it. At least once a week, I have a banana or two that is too brown and spotty for my picky family, but that is still good and firm. The peel goes into the compost heap, and the banana goes to the hens. It is usually gone within minutes.

- **Broccoli** – Broccoli is a delicious and nutritious treat for chickens. You can give it to them raw or cooked, as long as it doesn't have spices or oils on it. A great source of fiber, broccoli is high in protein and calcium. It is a low-calorie treat, but it is high in healthy nutrients, like vitamin C, vitamin K, and potassium.

Ones to Avoid

- **Potato Skins** – Many suburban chicken farmers offer their flock kitchen scraps—but avoid giving your chickens potato peels. Potato peels that have been exposed to sunlight and are starting to turn green are toxic. That green color comes from the chemical production of alkaloid solanine, which can sicken your birds. Sweet potatoes, on the other hand, do not produce this toxin in their skins; therefore, sweet potatoes and sweet potato skins are safe to give to your chickens.

- **Avocado** – The skins and pits of avocados contain a toxin called persin. Chickens, of course, would never be able to swallow an avocado pit, but it might be tempting to include avocado skins in a bowl of kitchen scraps for your hens. Be mindful of the food items that should be avoided to make sure your flock stays healthy.

- **Eggplant** – While the purple flesh of the eggplant is okay for chickens to consume, you should take care that your hens don't eat the green parts of the eggplant, like the stems of leaves. These plant parts contain solanine, a substance that is poisonous to animals. The eggplant, in fact, is a member of the nightshade family, which is known for having numerous poisonous family members.

- **Citrus** – Citrus fruits, like oranges, tangerines, lemons, and grapefruit, are not only loaded with sugar, but they are high in citric acid. You know how you get sores in your mouth if you eat too many oranges or tangerines? That happens to chickens, too, only they don't need to eat that much to have a negative reaction. The citric acid is too harsh for their digestive systems.

- **Onions and Garlic** – While onions and garlic are safe for your chickens to eat and won't cause them any harm, the strong flavor of onions and garlic may taint the taste of the eggs they lay. And not in a good way. Avoid giving your hens onions if you can. Small amounts may not impact the egg flavor, but why take a chance when there are plenty of other options available?

- **Rhubarb** – The anthraquinones in rhubarb are a natural laxative for chickens. Too much of it can give your hens a severe case of diarrhea that can lead to dehydration and death. And when the rhubarb plant is hit with a deep freeze, it produces high levels of oxalic acid, a substance that is deadly to chickens.

Greens to Try

Chickens instinctively forage. They wander around and try a bite or two of various plants that they find. Greens aid in digestion and can provide a punch of vitamins and minerals.

- **Vegetable Greens** – The rich nutrients in vegetable greens, such as kale, turnip greens, lettuce, cabbage, and chard, are great for chickens and can even result in richer eggs with deep orange yolks. They are loaded with vitamins A, C, and K, as well as calcium and iron. Many suburban chicken farmers add more greens to their flocks' diets in the wintertime to make up for the greens the birds are no longer getting by foraging. You can keep your hens happy, active, and engaged during the winter months by hanging a head of lettuce or cabbage in their coop. They can play with their food.

- **Lawn Greens** – Yes, your chickens will eat grass, but they are not grazers like sheep and cows. Yard grass is not tops on their list of favorite foods, so you don't have to worry about them decimating your lawn. If you have a yard that contains a mix of lawn plants and not a manicured, homogenous landscape, your chickens will explore it in search of things like clover and dandelions, two of their favorite lawn greens.

- **Garden Herbs** – Last spring, I planted several types of herbs in cute little containers and arranged them in a cluster on my deck, right off my kitchen. I envisioned myself stepping outside to snip some fresh herbs when cooking. That was a pipe dream. Less than a week later, I arrived home to find all my flower pots devoid of plant life, with the potting soil scattered about. My chickens discovered my little herb garden and completely destroyed it. Chickens love herbs. Mint, parsley, basil, lavender, and cilantro are among their favorites.

Greens to Avoid

- **Garden Greens** – The leaves of tomato plants, potato plants, and rhubarb all contain toxins that can make your chickens ill.

- **Yard and Landscape Greens** – Several common landscaping plants can be potentially dangerous if your chickens consume them. The bracken fern, for example, can poison your hens and cause them to suffer from weight loss, anemia, and tremors. The waxy leaves of azalea shrubs are also toxic to chickens and can cause weakness, muscle instability, and digestive issues. Holly leaves contain a substance called saponins, which, when ingested, will give your hens diarrhea and cause vomiting. Foxglove and rhododendron are also toxic. Oak leaves and the acorns from oak trees are also unsafe for chickens. Both contain dangerous levels of tannic acid, which cause kidney failure in chickens.

Growing Your Own Chicken Food

It might be tempting to cut costs by growing your own chicken food. While I always advocate for suburban chicken farmers to grow fresh greens, fruits, and vegetables that they can give to their hens as a treat, I advise people to use caution if they plan to grow their chickens' main food themselves.

As you recall from the earlier chart, chickens have specific nutritional requirements that change throughout their life span. The animal nutritionists at the feed manufacturing companies have used their scientific know-how to develop the best possible feed formulas to satisfy the nutritional needs of chickens during various stages of development. Unless you are an animal nutritionist who specializes in poultry, you probably won't be able to produce a well-balanced, comprehensive feed that is adequate for your flock.

Oyster Shells

Most commonly, suburban chicken farmers turn to oyster shells as a way to ensure their hens are consuming enough calcium. You can find sacks of crushed oyster shells at your local feed store. The cost is fairly low, and the oyster shells last a long time. Oyster shells are rich in calcium carbonate—just what your hens need. Chickens can dissolve the oyster shells in their digestive system, and the calcium is absorbed into the bird's body.

Other suburban chicken farmers may suggest that you save yourself the cost of a bag of oyster shells and, instead, crush up the eggshells from your own chickens and feed that back to them. Some folks swear by this, but I avoid doing it for two reasons. First, if my hens' eggshells are already so thin that I feel a calcium supplement is necessary, then the shells won't provide an adequate supply of calcium. Second, I don't want my hens to get a taste for their own eggshells. If they decide they like it, they may peck and eat their own eggs before I have a chance to collect them.

If you are looking for a way to recycle your eggshells and you want to stretch a bag of oyster shells, you could try mixing crushed eggshells with the oyster shells. That seems like a good compromise to me.

Photo Courtesy of Julia Cazier

Oyster Shells Versus Grit

Note that oyster shells are not the same thing as grit. Grit is necessary for digestion; therefore, the grit remains undissolved in the chicken's gizzard.

A Word About Lawn Care and Suburban Chickens

Homeowners living in the suburbs often have rules and expectations about lawn maintenance, either as part of their homeowners' association or simply from peer pressure. You don't want to be known as the house with the worst lawn in the neighborhood, am I right?

With these rules and expectations, however, come the lawn care service technicians who show up periodically to spray your yard so you can remain the envy of your neighbors and stay on the good side of the neighborhood association. But what are the lawn-care guys spraying on your grass? Is it safe for your chickens? Is there a way to have a good-looking yard while raising suburban chickens? Let's find out.

We will start with the last question first. Yes, it is definitely possible to keep your lawn looking nice while raising healthy chickens. But you may have to change your approach to lawn maintenance and adopt a more organic, chemical-free approach.

Things to Discuss with Your Lawn Care Service

If you use a lawn-care service, start by talking to someone there. Explain that you will be introducing backyard chickens to your home, and you want to find out what chemicals are being used on your yard. Fortunately, raising suburban chickens has become so popular that lawn care services know how to accommodate them. But here are a few things to ask.

Do you use synthetic fertilizers?

The answer you are looking for is no. Synthetic fertilizers contain harmful chemicals that can negatively impact your chickens. The chemicals can get into the birds' bodies when they eat the grass, or they can be absorbed into the chickens' bodies when they simply come into contact with the treated grass. Ask your lawn care service to use an organic fertilizer instead. There are several good ones on the market, so it shouldn't be too difficult to make the switch.

What precautions should I take with organic fertilizers?

Yes, organic fertilizers are best, but there are still some precautions you need to take. Organic fertilizers are made using the same chemical substances that are in synthetic fertilizers, only in their natural, non-man-made form. You should keep your chickens off the treated grass for at least 24 hours after the organic fertilizer has been applied and prevent your hens from dusting themselves in the dirt. The chemicals can absorb into their bodies. If you are concerned about the possibility of chemicals, albeit organic ones, in the eggs your hens lay, you may want to throw out the eggs for five or six days after a fertilizer treatment.

Do you use pellets or spray herbicides?

Part of the issue with weed killers is that they are easily absorbed. When your chickens come in contact with the herbicides on the grass or eat an insect that was in contact with it, that poison enters their bodies. Too much of it can lead to internal organ damage. Pellet herbicides are designed to last longer, so they pose a longer threat to your hens. Chickens may also think that the pellets look like a piece of their food or another tasty treat. Herbicides that are applied by spraying don't have the same staying power. They really only remain on the grass for 24 to 48 hours but are still a potential threat.

If weeds are a problem in your yard, you can try spot-treating just the weedy area. You can also opt for safer, organic weed killers or remove the weeds the old-fashioned way—pulling them by hand. You can also let your chickens do your weeding for you. Dandelions, clover, chickweed, and many of the other common plants that we consider to be weeds are favorites of chickens. If they have access to the weeds, they may not completely

eliminate them, but they can certainly help keep them from spreading and eat the seeds before they can spread.

Buying and Storing Chicken Feed

Farm and feed stores are probably your best source for purchasing chicken feed for your suburban chickens. It is possible to order feed online or buy it at a pet store, but you will find the best prices and selection at a farm store. The only issue is that the feed is sold in 40- to 50-pound bags. That's a lot of feed for four or five chickens. Let's talk about ways to make this work on a small scale.

HELPFUL TIP

Natural Pest Control

Chickens are great at eliminating pest problems in backyard gardens, but how can you keep them from eating your veggies along with the bugs? Experts suggest that allowing your chickens to forage for bugs during the autumn, after all of your crops have been harvested but before the ground freezes, is the easiest way to use your birds as pest control. Because chickens will eat adult bugs, eggs, and larvae, this will decrease your bug issue the following spring. In addition to this method, some owners allow their chickens to forage for bugs amidst crops chickens don't normally eat, such as potatoes, squash, or onions.

Share the Feed

A common practice among suburban chicken farmers, especially ones living in the same neighborhood, is to share the feed. One or more other chicken owners share the cost of the 50-pound sack of feed and divide it up equally. It is a good system that makes sense. The feed doesn't spoil before you can use it all, and you won't have to hassle with storing a large quantity of feed. Plus, it gives you an opportunity to see your fellow chicken friends every few weeks to swap feed and discuss chicken rearing. Believe me; it can be a hot topic of conversation.

Spoiled Chicken Feed

By design, chicken feed has been formulated to be fairly stable, but it can't last forever. In general, chicken feed should keep for about three months as long as it has been properly stored. Over time, the feed can go bad. It will smell terrible, be unappetizing for your flock, and may even make them sick. You can take a big step toward preventing your chicken feed from spoiling by making sure it doesn't sit around unused for a long time. If you only have a few hens, split a sack of feed with friends or neighbors who have chickens so the product is used more quickly.

Moisture and Humidity Trouble

Moisture and humidity are the two main challenges to chicken feed storage. The moisture can lead to fungal growth and turn the pellets and crumbles into a messy mush. At the manufacturing facility, the chicken feed is baked to remove excess moisture. But it can pick up unwanted moisture after that, at the manufacturer's plant, when it is shipped to the store, on the feed store shelves, and at your home. The risk increases if you live in a region with high humidity.

Feed should be stored in a cool, dry place. The container of feed—and we will discuss containers in a moment—should be placed out of direct sunlight. The heat of the sun can cause condensation to form on the inside of feed containers. Fungus likes humidity levels of 65 percent or higher, so the drier the storage location, the better.

Early on in our chicken-raising adventures, we built a lean-to-like addition on the side of our chicken coop that we planned to use as a feed room. It would be super handy, we thought, to have everything we needed right in one place. We quickly realized that this was not the ideal place to store our chicken feed. The sun beat down on this side of the coop all day long, and humidity levels exceeded 80% in the summer. We had created a petri dish for fungal growth. Oops. We also had a rodent problem.

Rodent Problems

Chickens love their grain-based feed, but they aren't the only ones. Mice, rats, and squirrels also love it. And they love that you have a huge supply available for them to eat whenever they want. Rodents are gross, and, I'll admit, I scream a little when I see one, but they can eat a tremendous  amount of your chicken feed. And then they invite their friends to come eat it too. Mice and rats also carry diseases that can get into the food and infect your birds. You need to make sure that both the spot you select for feed storage and the container you use are rodent-proof.

Insect Invasions

Like mice, insects can invade your chicken feed and cause problems. Yes, your hens like a tasty insect to snack on, but the bugs that sneak into your feed are usually not the ones they prefer. They are usually destructive, disease-carrying insects, like weevils and moths. To compound the problem, insects will lay their eggs in the feed containers, and the newly hatched bugs will have a ready source of food.

Feed Storage Containers

People storing a full 50-pound sack of chicken feed often use a metal drum or trash can. A benefit to this is that mice can't gnaw their way in. These containers are also waterproof. But the feed can react to the metal, so it is advisable to either line the trash can with a garbage bag or keep the feed in its original sack and place it in the can.

Plastic trash cans work too. In fact, this is what I use. Just my personal observation: I think there is less condensation with plastic versus metal trash cans, which is a plus for me. However, I have had to replace a plastic trash bin once because a determined rodent chewed its way in. And, more than once, a raccoon has been able to jimmy the lid off.

For smaller quantities of chicken feed, a plastic storage tote is a good option. I like the ones with the lids that snap closed with latches on either side. I'd like to see a raccoon figure out how to get into one of those!

One last word about storage containers and lids. Lids are only as good as the person closing them. Whenever one of my family members comes inside from feeding the chickens, I say, without fail, "Did you close the lid?" It has been several years, but I am finally getting them trained.

Free-Ranging and Foraging

> *My personal philosophy is that my chickens are happier and healthier with the opportunity to free range around my yard than they would be if they were locked up. During the day they are out and about. That is a decision for you to make.*
>
> KRISTEN HARRELL

We have an entire chapter coming up that is devoted to free-ranging your chickens, but we will touch on it briefly here as it relates to feeding your flock.

Chickens are natural foragers. They can keep themselves busy all day, going from place to place and searching out insects, seeds, and greens to nibble on. Long before chickens were domesticated, they managed to feed themselves solely on the morsels they found while free-ranging. As a species, chickens survived for thousands of years by employing this method. Does that mean you can skip the feed store and just let your birds feed themselves? Probably not. Here's why.

The wild chickens of the past did not have a livestock purpose. They were not kept for their eggs or meat. Based on the food they were eating and whether or not it met the minimum of their nutritional requirements, the chickens may not have produced as many eggs as today's suburban chickens. Nor did the birds live as long as they do today. Additionally, the quality of the eggs and meat most likely suffered.

Another key factor is the habitat in which the chickens forage. Wild chickens living hundreds of years ago foraged for food in a natural environment with no chemical herbicides or fertilizers, where only native plants and animals lived, and with a greater variety of plant life. The backyards of most suburban homes are a far cry from the wide-open prairies of the past. Modern suburban homes are landscaped using non-native plants, and the diversity of living things in this environment has been greatly reduced. Simply put, free-range chickens in today's suburban setting cannot adequately satisfy their nutritional requirements on the tidbits they forage. They still need to be fed a main diet of chicken feed but can be allowed to forage as well.

Chapter 5 Summary

We covered a lot of ground in this chapter, but proper nutrition is important. Now that you have a basic understanding of the nutritional needs of your suburban chickens, you can make informed decisions about the feed you give your birds, be it mash, crumbles, or pellets. In addition, you know what kinds of treats, including greens, fruits, and kitchen scraps, you can give to your hens and how best to store your chicken feed. You want your flock to be happy, healthy, and productive. The first step to reaching those goals is providing them with a balanced diet.

CHAPTER 6

Protecting Your Chickens from Predators

> *For predators, chickens are the bottom of the food chain—everything eats chickens! Also, chickens die ... sometimes for random reasons. I had one die and the only reason I can think of was that it was a Tuesday. The favorite chicken will always be the one a predator kills! You and your family will quickly learn the circle of life.*
>
> AMY JOHNSON

You may think that your chickens are more protected from predators in a suburban setting as opposed to a rural one, yet you still need to take steps to keep your flock safe from harm. After all, chickens are prey animals—plump, juicy, easy-to-catch prey animals. Their mere presence in your backyard can attract unwanted critters like raccoons and foxes.

Throughout this chapter, we will explain the common predators that plague suburban chicken farmers, strategies and tips for deterring predators, and your legal rights to protect your flock from predators. We will also discuss what to do and what not to do if a predator launches an attack on your henhouse.

I can tell you from personal experience that it is heartbreaking to go out to the coop in the morning only to discover that an animal has

decimated the flock. Our hens become like pets—like family—and losing one or more of them to a brutal attack is devastating. By taking some preventative measures and being diligent about backyard security, you can protect your chickens and your heart.

Common Predators in Suburban Areas

> *Our first predator issue was a dog that killed our first pair of birds. Since then I have had owls, coons, possums, minks or weasels, and foxes kill birds here in the past 15-plus years. A fence now protects my birds, and they are in a secure building at night.*
>
> CRAIG HANSEN

Many of the same predators that create problems for rural chicken farmers can also impact suburban flocks. As humans continue to encroach on the natural habitats of wild animals, the displaced wildlife will find ways to adapt. When we build more neighborhoods and subdivisions on land that was once an open field or a wooded area, we are taking away the homes of numerous species of wildlife. They have to go somewhere, so we should accept the fact that we coexist with wild animals. Of course, that doesn't mean we have to feed them our chickens.

Wildlife Predators

> *Rats want the chicken food, and raccoons want the chickens. You must be more diligent and fiercer about protecting your flock than they are. Encase your run with hardware cloth, dug six inches below the base. This prevents rats from digging under, which they are skilled at doing. Hardware cloth is the only material strong enough and with small enough openings to prevent rats and raccoons from getting through. Any material that raccoons can reach through is large enough for them to kill chickens. Unfortunately, chickens don't seem to be smart enough to avoid a raccoon's reach on their own.*
>
> ELIZABETH SORBY

Let's hear from our panel of experts on their experience with predatory attacks on their suburban chicken flocks and discuss some of the wildlife that you may encounter in your neighborhood.

Raccoons

Raccoons are the number one predator that visits my chicken coop. Raccoons are stealthy, clever, and fierce. They have little human-like hands that can turn doorknobs, lift window panes, and work to open simple locks and latches. Excellent climbers, raccoons can scale walls and fences with ease. They are opportunistic eaters. Raccoons will raid your trash can, empty a cooler full of food, and, sadly, break into your chicken coop.

Once inside, a single raccoon can kill your entire flock in just a few minutes. It will leave behind a scene of utter carnage. Blood, feathers, body parts—it is horrible.

I am not the only one to name raccoons as the biggest threat to suburban chickens. Kristen Harrell of Iowa notes, "Raccoons are my biggest predator. Locking up the coop at night is very important to keeping the chickens safe."

A raccoon doesn't need to completely breach your henhouse to cause devastation. If one can reach through the wire and snatch a hen, it will not let go. And it usually doesn't end well for the chicken. As Elizabeth Sorby of Washington explains, "Any opening that a raccoon can reach through is enough for it to kill a chicken. Unfortunately, chickens don't seem to be smart enough to avoid a raccoon's reach on their own." She added, "Sadly, I know from early experience that a raccoon can reach through a one-by-two-inch wire and pull out a chicken in pieces."

Hawks and Owls

Sometimes a chicken attack comes from above. Hawks, owls, and other birds of prey can swoop down, pluck a chicken, and be gone in a flash, leaving behind only a few telltale feathers. If your chicken run does not have a top enclosure or if you allow your hens to free range around your yard, they are vulnerable to an aerial assault.

Fortunately, most birds of prey, like owls, hunt mainly at night when your hens are safely tucked inside their coop, but they will also venture out at dusk. Hawks can be a problem throughout the day.

I have been lucky. Hawk attacks on my flock are fairly rare. I may lose one chicken every few years to a hawk or owl. I have even had one hen survive a hawk attack. My husband and I witnessed a young, overzealous red-tailed hawk dive down from the sky to snatch one of our hens, Chickeletta. But the brazen hawk totally miscalculated the size and weight of its prey. Chickeletta, a barred rock, is a healthy-sized, mature hen. The hawk couldn't lift her off the ground, try as he did. His struggle gave us time to react. And by react, I mean my husband grabbed a broom, and I grabbed a cowbell, and we ran toward the hawk yelling, screaming, and clanging the bell until the hawk gave up its would-be dinner and flew away.

Amy Johnson, who lives in Minnesota, has dealt with hawks as well as bald eagles. She says, "We lost a few hens to hawks. It didn't matter if we were in the yard; the hawks would swoop down and take a hen." Craig Hansen of Iowa has also encountered both owls and hawks. His solution is to keep them "in a secure building at night."

Coyotes and Foxes

Seeing coyotes and foxes in suburban neighborhoods is becoming more commonplace, which is causing some concerns for residents. Many people freak out at the idea of a coyote near their home, but often this is because folks think that coyotes are dangerous predators that will attack humans. That might be true of wolves, but coyotes almost never attack people. As the Humane Society of the United States notes, "More people are killed by errant golf balls and flying champagne corks each year than are bitten by coyotes." You are safe from coyotes; however, your chickens are not. "I've had one hen taken by a very wily coyote," says William Quigley of Washington.

Both coyotes and foxes are opportunistic hunters, but they are also somewhat lazy. They will go after easy prey, especially things that won't put up much of a fight. Coyotes and foxes are all about self-preservation. Rarely will they go after anything larger than they are.

Foxes and coyotes do most of their hunting at dusk, but they will also raid a henhouse in the dark of night. Cousins to the domestic dog, foxes and coyotes are skilled diggers. They will dig under a fence to get to your chickens.

Although raccoons are more of a problem where I live, we have had issues with both coyotes and foxes. I won't tell you about the time when I chased a coyote out of the chicken coop one night when I was home

alone, eight months pregnant, and wearing my slippers. But I will tell you that one summer, we lost several hens to a fox. We investigated and discovered that the fox had dug her den just behind our property in the corner of an open field. The opening of the den was adorned with feathers from our chickens. The fox had a litter of kits that she kept well-fed with our chickens.

Skunks, Opossums, and Minks

You might not think of skunks, minks, and opossums as predatory hunters, but they are. Like coyotes, they like to attack chickens because they are easy to catch, and the risk of injury from a chicken is minimal. When a skunk or mink kills a chicken, it will eat part of it right there and leave the rest of the body behind. You will notice bites out of the hen's breasts and abdomen, but the head and wings are untouched. Mink and weasels have a habit of killing several birds, taking a few bites out of each one, and stacking their bodies in a corner.

Rats

In urban and suburban areas, rats can be a real problem. "Rats want the chicken food," explains Elizabeth Sorby; however, they can also attack your chickens, causing injuries and death. "Rats can squeeze through

remarkably small openings," Sorby adds. Once inside, they usually head for the feed dish, but they will also kill young chickens or smaller bantams.

Rats might merely injure your chickens. If you find some of your birds with cuts and bruises on their feet and legs, it could be that a rat tried to pull them down from their roost.

Snakes

Like rats, snakes are primarily attracted to your chicken coop for the feed dish. But once there, they will take advantage of whatever food they can find. If you notice whole eggs are missing and there are no broken shells left behind, the culprit could be a snake.

Your full-sized adult hens are probably safe from becoming a snake snack, but chicks, young hens, and bantams may fall victim to a slithery predator.

Domestic Predators

Not all predators outside your henhouse are wild animals. Your chickens may be at risk for attacks by domestic predators. What do we mean by domestic predators? Dogs, cats, and humans, both kids and adults.

Dogs

Sure, dogs are man's best friend, but they are also predatory hunters. Even domesticated dogs still retain vestiges of their wild ancestors. While some breeds have a stronger prey instinct than others, all dogs—from the tiniest Pekinese to the mightiest Great Dane—have the potential to go after your chickens. "Our first predator issue was a dog that killed our first pair of birds," recalls Craig Hansen.

We have a dog—a very large dog. As a pup, he liked to chase our chickens, but he has never attacked them. But I know plenty of people who have had to reinforce their chicken run and add a dog run because their lovable, friendly pooch turned into a savage, chicken-killing machine every time they let it outside.

The most likely culprit of a dog attack on your flock, however, will not be your own pet but a stray dog or a dog belonging to one of your neighbors. A dog attack can happen overnight or during the day. If your birds were in their chicken run when the attack occurred, you will see signs of the fencing being torn down and ripped away.

WHAT TO DO IF YOUR NEIGHBOR'S DOG KILLS YOUR CHICKENS

One benefit to living in a suburban neighborhood is that you will have neighbors. One drawback to living in a suburban neighborhood is that you will have neighbors. And those neighbors may have dogs. Statistically, more than 40% of all American households include at least one furry, four-legged member. While I subscribe to the "dogs are man's best friend" philosophy, I know that this friendship doesn't always extend to chickens.

So, what do you do if you discover that your neighbor's dog has come into your yard and killed one or all of your suburban chickens? What recourse do you have? Here are some tips for dealing with this.

1. Talk to Your Neighbor

If you know who the dog's owner is, the first step would be to pay them a visit. I understand that you will be angry and heartbroken about the loss of your hens, but approach the conversation with your neighbor as calmly as possible. You don't want to make the situation worse by being confrontational, demanding, and aggressive. The neighbor may not even know what their dog has done and could be surprised and remorseful to learn the news.

Ideally, you have a good relationship with this neighbor already, and the neighbor is a responsible dog owner who will quickly offer to compensate you for the destruction their dog has caused and prevent future attacks. That would be the neighborly thing to do.

2. File a Police Report

Call the non-emergency number for your local law enforcement agency, explain what happened, and ask for an officer to come out and make a report. Many suburban chicken owners suggest you do this for every incident, even if the neighbor is remorseful and reimburses you for the loss of your birds. By having a police report on file, you are starting a paper trail. Let's hope the dog attack is a one-time event and you never lose another hen to a dog. But if the dog is a repeat offender, it will benefit you to be able to prove it with a string of police reports.

3. Ask the Officer to Speak to Your Neighbor

Again, if you are on friendly terms with the neighbor and they've made things right with you, this step might not be necessary. But if you had some pushback from the neighbor or they refuse to acknowledge the incident, ask the responding police officer to speak to them. The officer doesn't need to be threatening or intimidating. Just the uniform and badge are often enough to make your neighbor take notice and control their dog better.

4. Check Your State and Local Laws and Ordinances

There are laws and ordinances in every county and municipality in the country covering animal ownership, and these have sections regarding nuisance dogs. In many places, the nuisance dog section contains ordinances addressing instances between dogs and livestock animals kept by neighbors. This includes chickens, even suburban chickens. The law will outline the recourse you have, the compensation you are due, and the measures you can take against the offending dog and its owner. If your local and county laws don't cover dog-on-chicken attacks, your state's laws do. In fact, a quick Google search is all that is needed to find these.

Just as an example, the state of Michigan's law regarding dogs and chickens states, "Any person, including a law enforcement officer, may kill any dog which he sees in the act of pursuing, worrying or wounding any livestock or poultry or attacking persons, and there shall be no liability on such person in damages, or otherwise, for such killing." This seems harsh. I would never advocate for killing a dog, even one that has attacked my hens, but it shows you that the court system recognizes the seriousness of domestic dog attacks on suburban chicken farmers.

5. Alert Animal Control

Contact your local or county animal control department. Explain the incident to the animal control officer and ask if the offending dog has had any previous complaints. You will also be able to find out if the offending dog has a dog license. In some areas, an unlicensed dog is viewed as a public nuisance and can be picked up by animal control.

6. File a Lawsuit

If you cannot work with your neighbor to resolve the problem to your satisfaction, if the attacks continue, and if the police and animal control officers have not been able to protect your hens, you may consider filing a lawsuit against your neighbor. This should be a last resort after you have exhausted other efforts to resolve the conflict. However, you are within your legal rights to protect your suburban chickens from nuisance dogs and to seek rightful compensation.

Cats

Cats, even domestic cats, are excellent hunters. However, they generally do not attack prey items that are larger than they are. My own cat, a stealthy huntress, is an average-sized house cat. She is not larger than our hens and probably weighs about as much as the adult chickens. She tends to ignore our chickens; however, she does like to hunt the field mice that try to nip the chicken feed.

A feral or domestic cat may go after smaller chickens, young chickens, or bantams. Cats won't eat the wings and feet of the chickens they kill and will leave the carcass behind. If you find a slaughtered hen with its wings and feet intact, that is a sign that a cat may be to blame.

Reinforce your chicken run and henhouse to keep feral or domestic cats out. Unlike dogs, cats won't dig under a fence. Instead, they will squirm through small openings or climb over fences.

Humans – Adults and Kids

It is rare, but occasionally a person will sneak into a backyard and steal a suburban chicken or two. In fact, reports of chicken thefts have increased in recent years. I can only speculate as to the reasons, but perhaps some of it has to do with the rising egg prices. Cases of adults killing backyard chickens are even lower, and sadly, most of the offenders were people struggling with mental health issues.

The children in your neighborhood could also be a threat to your suburban flock. While it can be argued that some neighborhood kids are just rotten, in my experience, it has been curious, unknowing children that posed a problem.

We diligently close our coop every night at dusk. One summer night several years ago, the neighbor's ten-year-old son invited a few buddies over for a sleepover in a tent pitched in their backyard. As the sun set, the neighbor boy decided he wanted to show off our hens to his friends, but the hens were already locked in for the night. So, the boys crept into our yard and opened the coop. The hens were roosting and didn't want to come out to play, so the boys entered the coop and ended up chasing the squawking chickens around. The noise alerted the dog, who alerted us. No birds were harmed in this story, but it had the potential to be detrimental to our flock.

A Mystery at the Henhouse: Who Killed the Chickens?	
The Clues	**Possible Suspect(s)**
One bird missing , scattered feathers	coyote, hawk, fox
Multiple birds killed, heads missing, breasts & necks torn open	raccoon
One bird killed, only breasts, thighs, & abdomen eaten	opossum
Deep cuts to the head & neck	owl
One bird killed, wings & feet not eaten	domestic cat
Several birds killed but not eaten	domestic dog
Young chickens killed, abdomen eaten, foul odor	skunk
Multiple birds killed, bite marks to the neck, heads eaten, corpses neatly piled	mink
Multiple birds killed, bite marks to the neck, heads eaten, corpses neatly piled, foul odor	weasel
Young chicken missing	snake
Multiple birds killed, rear ends mutilated	marten
Digging under the fence	fox, coyote, raccoon, domestic dog
Young hen missing	rat
Fence destroyed	domestic dog
Head bitten off, body partially buried	bobcat
Latch opened	raccoon, human

Had we not discovered the breaking and entering in progress, the boys may have succeeded in letting our chickens out of the coop. I doubt they would have bothered to round them back up and secure them in the coop when they were done playing show-and-tell. The chickens could have been left outside overnight, putting them at risk of a predator attack.

Deterring Predators

The best way to keep your suburban chickens safe from predators is to create a home for your birds that cannot be breached by a hungry animal looking for an easy dinner. Predators are crafty and clever, so this is not as easy as it sounds. "You have to be more diligent and fiercer in protecting your flock than you think you'll need to be," explains Elizabeth Sorby. The following are tips and tricks you can try to deter predators from reaching your hens.

You should inspect your chicken housing, fencing, and latches on a regular basis. Make any necessary repairs as quickly as possible. Look for signs that a predator has tried to get in and reinforce those areas. Outsmarting predators is an ongoing challenge. You can't afford to become complacent, for your hens' sake.

Fencing

If you are lucky enough to live in a housing community that allows you to have privacy fences, you have two lines of defense against predatory attacks in your backyard—your privacy fence and the fencing used in the chicken run. For our purposes, we will focus only on chicken-run fencing.

Chicken wire, with its iconic hexagonal pattern, may work to protect your tomatoes from foraging hens, but it is not effective at keeping predators out. Raccoons, dogs, and coyotes can easily bend this wire. Rats and snakes can go through the holes.

A much better choice is half-inch welded wire, also known as hardware cloth. It is a much more durable material that is impervious to foxes, raccoons, and coyotes. The gauge is small enough to keep out snakes, mice, and weasels.

"The number one thing you can do to protect your flock is to completely encase your run in hardware cloth, including digging it down six inches or so around the base," says Elizabeth Sorby. "Hardware cloth is the only material strong enough and with small enough openings."

William Quigley of Washington agrees. He says, "Having an enclosed run with the fencing buried six inches deep helps." Foxes, coyotes, and dogs will all tunnel under a fence but will be deterred from digging if you extend the fence underground.

In addition to digging, foxes, coyotes, and some dogs can easily jump over a five-foot-tall fence. You could install a six-foot-high fence or try

another trick. Run a strong, thick wire along the top of your fence and thread a long section of PVC piping through it. When the fox or coyote tries to scale the fence, the PVC pipe will turn. The predator can't get a foothold to get over the fence.

The PVC trick also works to discourage hens from roosting near fences, where they are in reach of raccoons. Cut a small section of PVC and put it loosely around the roost. Tack a nail in the roost to keep the PVC where you want it to stay. When a hen tries to roost on it, the PVC will turn and spin, knocking her off balance. She won't like it, so she will move to a different spot.

Netting

Poultry netting, rolled netting made from strong, durable plastic, is flexible and lightweight enough to stretch across the top of a fenced-in chicken yard. It works well to keep your chickens from flying out and is effective at preventing hawks from diving down and swooping up one of your hens. However, poultry netting is not predator-proof. It won't keep weasels or raccoons out.

Alarms

There are a number of products on the market that are designed to help keep your chickens safe from predators. One such product is a coop alarm. You can get both motion sensor and sound sensor alarm systems. Each will detect the presence of a prowler outside your coop and emit a loud noise to scare it off.

A drawback to this type of coop alarm is that your neighbors may complain if the alarm is frequently set off, especially if it is triggered by a rabbit, squirrel, or other non-threat animal.

Motion Lights

Another type of coop alarm uses bright lights, not sound, to scare off the predator. Although there are specific chicken coop motion lights available, you can also consider using regular motion-sensor yard security lights.

William Quigley offers up an ingenious suggestion. "I use glow-in-the-dark tape to put two owl eyes on my coop door. The hens never see it because it's only visible when the door is closed. Does it work? Maybe ... it looks plenty creepy to me!"

Locks and Latches

As we discussed earlier when we covered housing for your suburban chickens, all the doors, windows, and other access points must be able to be completely closed and secured with a lock or a latch. Predators are pretty smart. As a rule of thumb, if a five-year-old child can figure out how to open the latch or lock, then a raccoon can too. Multi-step locks and latches should do the trick.

Can I Trap or Kill a Wild Predator?

When you have an ongoing predator problem, it may be tempting to want to shoot or trap the offending wild animal. Before you load a shotgun or set out a trap, you need to be aware that the laws regarding killing and trapping wild animals, even nuisance animals that are killing your chickens, vary greatly from state to state, county by county, and animal by animal. If you violate these laws, you could face stiff fines and criminal charges. Even using a live trap to relocate a predator could be against the law, depending on where you live.

For lawmakers, there is a delicate balance between protecting livestock animals and protecting wildlife. All states recognize the rights of poultry owners to protect their flocks. They just want to make sure it is done ethically, humanely, and in accordance with the law.

If you feel it is in the best interest of your suburban flock for the nuisance predator to be killed or trapped, you must start by contacting

your local wildlife service department, your local US Department of Agriculture office, or the local office of the Department of Natural Resources. The trained individuals working there will be able to answer your questions, explain the legalities for your area, and offer resources. Someone from one of these agencies may even come to set traps at your house, or they may tell you that you can legally trap the predator yourself. Either way, it is important to do your research so your activities stay on the right side of the law.

HELPFUL TIP

Keep it Trim

Predators looking for an easy meal are abundant, even in a suburban farming environment. So, in addition to providing a secure and reinforced coop, it's a good idea to keep vegetation trimmed near the coop to eliminate potential hiding places for your local predators. Overhanging tree branches may also provide an easy access point for predators looking for a chicken dinner.

When an Attack Happens

Ideally, that header should say "If an Attack Happens," but unfortunately, dealing with predators is a fact of life for suburban chicken farmers. When a wild animal does breach your chicken coop defenses, what can you do?

Responding to an Attack in Progress

Let's say your chicken coop alarm starts going off, or your dog starts barking, or you hear a lot of agitated squawking coming from your henhouse. This could mean that an attack is occurring. What do you do?

The first thing you should do is try to chase off the predator. The majority of the wild animals that you encounter in a suburban setting do not pose much of a threat to humans. You don't have to worry about a fox turning on you. Most of these predators are just looking for an easy meal. And they are all about self-preservation. They will run from a

fight with something larger than they are because they don't want to get injured or killed. With that said, you can probably run out to your coop, screaming and yelling, without putting yourself in danger.

Wild animals don't like loud, sudden noises. Your yelling might be enough to scare off a predator. We keep a small cowbell on a hook next to the back door. If a predator comes close, we clang the bell. It is loud and annoying and does the trick. It will send a coyote or raccoon running.

I have used a broom to shoo off a predator on more than one occasion. I have also thrown rocks, but my aim is not the best. A fellow chicken friend of mine fills a Super Soaker with ammonia. When she catches a raccoon near her coop, she blasts it with the strong-smelling liquid.

I have a large-breed dog, and I'm not afraid to use him! He can be awfully territorial about his yard, so he naturally wants to chase away any animal that doesn't belong there. He has scared several animals away from the chicken coop, earning himself a treat and ear scratches for his good-boy deeds. He has even tussled with a few raccoons. When a neighbor worried about this, I reminded him that my dog is routinely vaccinated against rabies. I am not!

Assessing Injuries

Many attacks on chicken coops don't have happy endings. If you stop an attack in progress or the predator gets scared off before it finishes the job, you may find frightened and injured birds rather than corpses.

Assess the injuries to your hens. If the bird is alert and active and the wound is not too severe, you may be able to provide first aid to help her heal. Until the wounds have completely healed, you should isolate the chicken from the others in the flock. Chickens are attracted to the color red and will want to peck at the hen's injury site, making it slower to heal.

More traumatic injuries will require professional care. Does your local veterinarian offer poultry services? Is there an animal emergency clinic nearby? Be warned that emergency medical treatment for animals, even chickens, can be costly. Euthanasia, as heartbreaking as that is, might be the best option for a severely injured chicken.

Regrouping After an Attack

An attack on your henhouse is a wake-up call for you to reinforce your chicken coop and chicken run. Add more latches. Repair fencing. Fill in holes. Nail down loose boards. Look for every possible access point and double down on your defenses. If a predator was successful in finding an easy meal at your coop, it will remember that and come back. Be ready for its return and take steps to thwart its efforts. Keeping predators at bay is an ongoing battle, but you can be victorious.

Photo Courtesy of Julia Cazier

Chapter 6 Summary

Sadly, there are plenty of animals out there that would love to get a taste of your chickens. Even in the suburbs, you will find wild animals like raccoons, foxes, hawks, snakes, and coyotes that will enter your neighborhood and try to sneak into your backyard. Some threats to your chickens even come from within your neighborhood. Your neighbor's pet dog might launch a surprise attack on your flock with devastating results. When a wild animal kills a chicken, there is no one to blame but the wild animal. When a pet dog kills a chicken, however, the animal's owner bears the responsibility, as we discussed in this chapter.

Ultimately, keeping your flock of suburban chickens safe from predatory attacks comes down to how predator-proof your fencing and housing are. Losing a chicken or two to a sudden animal attack is devastating, but unfortunately, it is something that comes with keeping chickens. By reinforcing your chicken run and chicken coop, you are taking important steps to ensure your flock's safety and security.

CHAPTER 7

Urban Chicken Ordinances

You may have decided that you really want to raise your own chickens, but are you allowed to? For people living in rural places, that's not an issue. There are few if any, restrictions on chicken farming in the country. Unfortunately, that is not the case for folks living in the suburbs. Before you invest in a henhouse, chicken run, and baby chicks, you need to make sure that your new hobby won't land you in legal trouble.

In this chapter, we will address the legalities of keeping chickens in a suburban setting. Even though laws, rules, and ordinances for chicken farming vary greatly from neighborhood to neighborhood, city to city, and state to state, we will strive to give you an overview of chicken laws. We will try to point you in the right direction so you can find out the laws and ordinances in your particular area. We will cover your right to farm, as well as your rights and responsibilities as a member of your housing community. If urban chicken farming is not yet permitted in your community, we will offer suggestions for you to advocate for change.

Understanding Laws and Regulations

We aren't all lawyers, so understanding complex legislation can be challenging. To make it even more complicated, the rules change depending on where you live. Your state, county, city, and even your housing association can all make rules that will dictate whether or not you can start a suburban chicken flock.

Right to Farm

Every state in the US has a set of Right to Farm laws. Although they all differ slightly, they are basically designed to protect individuals living in rural areas from nuisance lawsuits that their neighbors may file against them. Right to Farm laws arose to address an influx of lawsuits that were being filed by people moving from cities or suburbs to rural areas who were annoyed that their neighbors were raising livestock. Right to Farm laws protect the rights of farmers, ranchers, and homesteaders to do what most people living in rural areas do—keep livestock animals. The laws recognize that there will be noises and smells as a result of rural agriculture; therefore, a farmer cannot be sued by a neighbor complaining about the stench.

Legal Terms

Right to Farm laws, however, don't generally extend to suburban chicken farmers. Although these are state laws, local laws can supersede them. Locally, suburban chicken farmers are impacted by laws, regulations, ordinances, and acts and statutes. But what do these mean, and how do they differ? Here is a brief rundown.

What Is a Law?

A law is really a catch-all term for rules or collections of rules that are intended to control and dictate actions and conduct. Laws are enforced by officers at the federal, state, county, and municipality levels.

What Is a Regulation?

Regulations may carry the same weight as a law, but they are more like official rules. Regulations are issued by authoritative agencies that oversee specific areas. Although they are administrative agencies or organizations, they are able to establish and enforce rules that impact their area. For example, a state's department of education may create regulations that schools must abide by.

What Is an Ordinance?

Ordinances are local laws that have been enacted by the governing body of a city, town, or municipality. Most ordinances focus on things like the upkeep and safety of the community. Local ordinances prohibit littering and graffiti. They set housing codes, public parking rules, and safety standards. Ordinances are enacted regarding land use and zoning too.

What Are Acts and Statutes?

Even though the names are different, acts and statutes are essentially the same thing. Acts and statutes are laws passed by a legislative body, like the federal government or a state government. You have heard of the Affordable Care Act. That's an example of a federal act. At the state level, acts or statutes might address water usage in the state or provide for protection for wildlife.

In the vast majority of places, it will be local ordinances that will dictate whether you are allowed to keep chickens or not. Although you own your property, your city or town will have zoning laws that restrict how you can use your land. But there is one more organization that can also impose rules and restrictions on what you do on your own property—the homeowners' association.

What Is a Homeowners' Association?

Homeowners' associations, also called property owners' associations, are private organizations that regulate the do's and don'ts for people living in a subdivision or community. Suburban housing communities are created by real estate developers who purchase a large plot of land and subdivide it into smaller lots, then oversee the construction of residential houses on those lots. When someone purchases one of those homes, they agree to abide by the rules set forth by the homeowners' association.

The purpose of homeowners' associations is to maintain the high standards and appeal of the housing community. They want to make sure that everyone living there does their part to keep the community looking nice for the other residents and future homeowners. The rules a homeowners' association lays down may restrict the paint color on the exterior of your house, where you can install a privacy fence, and whether or not you can operate a business from your home. It is not a legal governing authority, but the homeowners' association can restrict homeowners in its neighborhood from raising chickens. Or they can give the green light to your suburban chickens.

Concerns Addressed by Regulations

Rulemaking entities, like municipalities and homeowners' associations, don't want to keep you away from fresh eggs or stop you from having a new hobby. They are concerned about other things, like noise, odors, and safety issues. Let's look at some of the common areas of concern that often pop up in ordinances and regulations.

Number of Chickens

In suburban areas where backyard chickens are allowed, you can expect there to be ordinances regulating the size of the flocks. These are meant to prohibit overzealous chicken owners from filling their backyards with dozens of birds. In general, suburban chicken flocks are limited to between four and six birds.

Banning Roosters

We have an upcoming chapter covering the pros and cons of roosters, but suburban chicken farmers may find that it is a moot point because their town prohibits roosters. Roosters are loud and aggressive. They crow early in the morning and throughout the day. They are protective over their hens and may act  aggressively toward humans. In a suburban setting, it is probably a good idea to raise only hens and forgo the rooster. And, yes, you will still get eggs from your hens without a rooster. Read Chapter 9 to find out why.

No-Slaughter Rule

In suburban settings, you may find there are rules preventing backyard chicken farmers from slaughtering their chickens for meat. Most suburban chicken farmers are in it for the eggs, so they are not bothered by this no-kill rule.

No Free-Range Rule

In Chapter 8, we will outline the pros and cons of free-ranging your chickens; however, many communities have no free-range rules. In these areas, you will be required to always confine your hens to a chicken run and chicken coop.

Nuisance Complaints

In areas where suburban chickens are permitted, there will probably be clauses written in the regulations to address nuisances. These clauses encourage chicken owners to maintain their backyard coop, keep the area clean and odor free, properly dispose of soiled bedding materials, and take steps to discourage rodents and other pests from coming into the neighborhood. Nuisance clauses outline the expectations, as well as the repercussions for suburban chicken farmers who do not properly care for their flock.

Coop Placement Requirements

The ordinances and regulations will likely address the placement of chicken coops and outdoor runs with setback requirements. Suburban chicken farmers will need to keep their henhouses a certain distance from neighboring homes and property lines. The specifics of coop placement requirements will differ depending on the lot size in the community.

Housing Requirements

Suburban chicken farmers may also find requirements for the type of coop, size, construction materials, and fencing materials within the ordinances. The rules are intended to make sure the birds are housed in an adequately sized coop that will protect them but one that is also attractive enough that it is not viewed as an eyesore. If you plan to build your own coop, you may need to get a building permit and have your coop design preapproved. "When we first got our chicken permit, we had to have our coop inspected," recalls Amy Johnson of Minnesota.

Fees and Permits

Depending on the town or municipality, regulations or ordinances may even state that suburban chicken farmers must get a permit, pay a fee, and register their flock with the town. Fees, if required, are typically nominal. Amy Johnson explains that in her Minnesota community, "We have to have a chicken permit, and we aren't allowed to have a rooster. We also need to place our dirty chicken bedding and poop in bags that we put in the trash."

Health Codes

Some ordinances may address the health of suburban chickens. Health codes will require suburban chicken owners to contact the local health department and take other precautionary steps if they believe their flock may be infected with avian influenza or another communicable disease.

Finding Out the Rules and Ordinances in Your Area

> 66
>
> *We must have a chicken permit and are not allowed to have a rooster. We also need to place our dirty chicken bedding and poop in bags and put them in the trash. When we first got our permit, we also needed to have our coop inspected.*
>
> AMY JOHNSON
>
> 99

Since there can be more than one agency that has a say in what you do in your backyard, it may be confusing to know where to start to find out the rules and regulations in your area. Where do you start?

Start with your homeowners' association if you have one. Ask if backyard chickens are permitted in your neighborhood and, if so, find out all the rules and requirements. There is probably a flyer or handout that details all of these so that you have it in writing and can refer back to it.

If you don't have a home-owners' association, contact your city hall or local zoning office. The zoning office will be able to provide you with ordinances for zoning and land use that pertain to your address. A city official at your city hall can inform you of the local ordinances regarding backyard chickens. If they don't have a list of restrictive clauses to give you, ask about flock size, housing requirements, and whether you

HELPFUL TIP

Community Engagement

Local ordinances for urban chicken farming vary, sometimes wildly, from county to county. Joining a local chicken-keeping group to connect with other enthusiasts is an excellent way to share knowledge and learn from the experiences of other farmers in your local area. Local groups can be found through social media platforms, such as Facebook or Meetup, or meetups listed in your local newspaper.

can have a rooster. Ask if there are any other restrictions or rules that you need to know. For example, in some areas, you may be prohibited from raising backyard chickens for financial gain, meaning you won't be allowed to sell the eggs.

It might be a good idea to also speak to an officer at your local animal control department, as well as someone with your local building department, to find out if either of these departments has rules or parameters you need to follow.

Inform Your Neighbors

> 66
>
> *My neighbor hates chickens. She is afraid of them. When she moved in, we put up a fence. I give her eggs, and she and my chickens live in harmony together. Honestly, with a six-foot-high privacy fence, no one knows we have chickens unless we tell them. A city bus or a barking dog is louder than the hens are. If you have a crabby neighbor or get a complaint, figure out how to accommodate the situation. A six-foot-high privacy fence easily fixed the situation for us. Offer eggs ... everyone loves eggs.*
>
> AMY JOHNSON
>
> 99

When you make the decision to get backyard chickens, you should have a conversation with the neighbors whose yards adjoin yours. Inform them about your plans and assure them that you will be setting up your suburban chicken coop to the letter of the law and will strive to keep your chickens from causing them any stress. Answer all of their questions to the best of your ability. Listen to their concerns and assure them that you will address those.

Ideally, your neighbors will be supportive of your new hobby or, at least, be ambivalent about it. If, however, you get some pushback from a neighbor, that doesn't mean you are out of the chicken business. Stay calm and friendly, but let them know that you have the right, under the

law, to raise chickens in your backyard and plan to exercise that right. Ask the neighbor to let you know if your birds bother them in any way. Then do everything in your power to make sure your hens do not become a nuisance. Once your hens begin laying, it might be the neighborly thing to give your neighbor a dozen eggs from time to time.

What if Suburban Chickens Are Not Permitted in My Area?

Backyard chicken raising is a growing hobby, and many communities have been revisiting old rules that prohibited poultry within city limits. Fortunately, most places are more open to the idea than ever before. If backyard chickens are banned in your community, what can you do?

Follow the Law

First, don't break the law. Don't think no one will know. They will. If you are caught keeping chickens in a prohibited area, you will face fines, and your birds will be confiscated. You could face a lawsuit and court costs. It is not worth it.

"You must follow the laws in your city/area about how many chickens you are allowed and whether you can have roosters," explains Elizabeth Sorby. "You can go to the local government website and type 'chickens' in the search bar."

Recruit Others to Your Cause

Seek out other chicken enthusiasts in your area. You will be better able to get the attention of your city council or homeowners' association board if there are a number of people who support your idea.

Gather Facts

Next, do your homework. Know the benefits of backyard chickens, as well as the drawbacks and complaints. Make a list of all possible concerns that may arise, and prepare factual information for each one. Find experts that can speak to issues like noise, smell, health risks, and so on. Contact your local agricultural extension office. If they don't have a poultry expert on staff, they can certainly direct you to one.

Find Support Online

Find support online by creating a grassroots social media organization. A Facebook page for your cause or information on other social media platforms will give you a vehicle for connecting with other people and disseminating information. You can also use this page to network with other groups around your state or region that are also working to change the suburban chicken laws in their area. You can share resources and ideas. The more likes and engagement you get on your social media pages, the more impressive your movement will be when you present it to your city council.

Write a Proposal

Make a detailed proposal for your city council that is clear, concise, and specific. Include the restrictions that you believe are fair, such as a limit of five hens, no roosters, no free-ranging, no backyard slaughtering, etc. Of course, all of these points will be debated and discussed, but by including them in your initial proposal, the city council members can quickly see that you are requesting approval for small-scale operations.

Photo Courtesy of Mary Eaton

Submit your proposal to your local city council or governing body and request that the proposal be discussed at the next board meeting. Plan to attend that meeting and bring a group of supporters with you. Be respectful, informed, and articulate when you present your proposal, but keep in mind that you will probably not get approval right away. The board will want to do their own research before they put the issue to a vote. They will want to hear from residents who oppose the issue and listen to their concerns. If you have pulled together your research, you will already have rebuttals to these concerns. Still, it may take months, even years, before you can convince the council to change the law regarding suburban chickens in your area.

Be Patient

It took, for example, more than three years before the backyard chicken ordinance was finally passed in South Bend, Indiana, in 2013. The city council debated the issue and heard arguments from residents for and against the proposal. In the end, they set a limit of six hens and no roosters, required that coops be at least 15 feet from all property lines,

banned on-site butchering, and prohibited commercial operations. There is also a $20 annual chicken permit fee.

In Nashville, Tennessee's Davidson County, the previous ordinance prohibited domesticated farm animals from residentially zoned districts but has since been revised to permit backyard chickens. Chicken owners are required to get a permit from Metro Health Animal Control, and roosters are not allowed.

The ordinances are different for people living in suburban Savannah, Georgia, and in suburban areas outside the Savannah city limits in Chatham County. Within the city limits, suburban flocks are capped at four hens. Roosters are not allowed, and permits are not required. In Chatham County, homeowners can have one chicken per 1000 square feet of high-ground property, a permit is required, and the coop must be inspected. The chickens must be kept for personal use only, and commercial operations are prohibited.

When a group of chicken enthusiasts sought to raise hens in Cedar Rapids, Iowa, they discovered that backyard chickens were prohibited within the city limits. A grassroots group was organized, and the members contacted the Cedar Rapids city officials to discuss changing the law.

The group worked closely with officials to write an ordinance that would be fair to all residents. In addition to flock size, coop size, permits, and enclosure requirements, suburban chicken farmers in Cedar Rapids are required to notify their neighbors about their plans to raise chickens. They must also take a two-hour class on the challenges and obstacles of raising chickens in a suburban setting. The Cedar Rapids city council approved the new ordinance in 2010.

Chapter 7 Summary

Raising suburban chickens is a rewarding hobby, but it is not legal in some places. To avoid breaking any laws—and to prevent fines, lawsuits, and the forfeiture of your hens—you need to abide by the laws, regulations, and ordinances in your community. Local laws pertaining to backyard chickens supersede state and federal laws. Even if your city allows backyard chickens, your homeowners' association may prohibit it.

Contact your homeowners' association, city hall, local animal control department, and others to help you understand the laws and ordinances that apply to your address. Understand all the restrictions and rules, such as the number of chickens you can legally keep, the placement of the coop, and how soiled bedding is to be discarded.

With the increased popularity of backyard chickens, more municipalities are willing to listen to residents who want to join this trend. It is possible to work with other like-minded chicken lovers and your city officials to work out compromises that will change existing laws and permit homeowners to keep backyard chickens, providing they follow the rules and take care that their birds do not become a nuisance and an annoyance to their neighbors.

CHAPTER 8

Free-Ranging and Foraging

C hickens are natural foragers. In the wild, chickens spend their days moseying around, searching for insects, and basically just puttering around. Domestic chickens still have this instinct but often can't exercise it because of their circumstances. That creates some unhappy hens, which in turn, impacts the size and quality of the eggs they lay. Commercial egg producers figured this out. In recent decades, they have realized that happy hens are more productive and give better-quality

Photo Courtesy of
Anne Greenwood

eggs; therefore, more and more commercial egg producers are opting for free-range, cage-free poultry practices, which they brag about on the egg carton labels.

Photo Courtesy of Kelly Kneeland

Free-ranging doesn't necessarily mean that you are letting your chickens run wild. I learned this when the neighborhood supermom called me a free-range mom. I took it as an insult, but after a few Google searches, I realized that free-ranging is a good thing—for chickens and for children. In this chapter, however, we will focus only on the chickens.

We will discuss the benefits of free-ranging your suburban chickens and how to wrangle your birds back to the coop. We will also include tips for keeping your free-range hens safe, as well as other drawbacks and obstacles to free-ranging.

Why Free-Range?

You will always find a backyard chicken farmer telling you that free-ranging is the best option for your hens, but why is that? While it is true that the chickens will be happier if they have a chance to stretch their legs—and of course, the mental health of your birds is important—there are plenty of other benefits to letting your birds out of the chicken run so they can roam free. Let's look at some of these benefits.

1. A More Varied Diet

When your hens are confined to a coop or run, they can really only eat what you offer them. Although their commercial chicken feed is formulated to satisfy their nutritional needs, who wants to eat the same thing day after day? Wandering hens have the opportunity to select their

own meals from the insects and plants they encounter. It livens up their boring diet—literally. Chickens love to catch and eat live insects.

2. Better Egg Quality

Because of their varied diet, free-range hens lay richer-tasting eggs. Crack open a store-bought egg, and you will find a pale yellow yolk. Crack open a free-range egg, however, and the yolk will be a bright orangey-yellow. The vitamins and minerals from the diverse foods they ingest trickle down into the eggs the hens lay. And to you, when you eat the eggs.

3. Your Chickens Will Be More Active

Being cooped up all day in a coop or enclosure, your hens don't have the opportunity to exercise their bodies. When you open up the coop door and allow them to free-range, the birds will enjoy the activity they need to keep them healthy. After all, everyone can benefit from adding more movement to their day, including hens. As a result, they will be stronger and healthier.

4. Tastier, Healthier Meat

As a suburban chicken farmer, you probably aren't raising your birds for their meat, but if you are, you will notice the difference that free-ranging makes. The increase in activity and the natural food the hens forage translates into tastier meat that is all-natural and healthier than the poultry you find in the supermarket.

5. A Cleaner Coop

The more your chickens are confined to their coop, the messier it will get. Your coop and chicken run will stay cleaner longer if the birds are allowed to free-range around your yard from time to time. Of course, that just means they are doing their business elsewhere, as we will discuss momentarily.

6. Cheap Entertainment

It sounds crazy, I know, but watching your chickens free-range around the yard can be quite entertaining. My husband brings out a lawn chair and will sit and watch "his ladies" for hours. This is when you can really learn the individual personalities of the hens and understand their flock dynamics. You'll see drama, conflict, loyalty, and intrigue. It's better than anything on TV.

7. Reduced Feed Costs

I don't suggest that you ever try to cut corners when feeding your suburban chickens. Their health and nutritional intake are too important. You should view your chickens' foraging activities as a supplemental feed source. However, you may notice that your hens are eating less of their pellets on the days that you allow them to free range. They are filling up on delicious (to them!) bugs, green sprouts, seeds, and fruits, all of which are loaded with vitamins and minerals.

8. Insect Control

Foraging hens will devour insects, including flies, grasshoppers, ticks, mosquitos, and bees. They love worms, grubs, and caterpillars. Let your suburban chickens forage through your flower beds and vegetable garden—with supervision, of course—and they

Photo Courtesy of Kristen Harrell

will help you out by eating the insects that are eating your plants. It is a natural form of pest control. As an added bonus, the chickens will poop in your garden, leaving their nitrogen-rich droppings to fertilize the soil.

9. Less Bullying

In an effort to establish their cliquish hierarchies, hens will sometimes pick on other hens. The bullying gets worse when the birds are confined together all the time. Remember how irritated you got with your family when you were all in COVID lockdown? It is like that for your suburban chickens too. When they free-range, it gives them an opportunity to get out and away from some of their coop-mates, if only for a while. Bullying gets worse when hens are bored, so the mental stimulation that comes with free-ranging will help reduce snippiness.

10. Mental Stimulation

Happy hens are ones that have plenty of mental stimulation. Free-ranging can provide that for your flock. When they step out of their coop, they can see, hear, taste, and smell new things. They can try out their problem-solving skills and stretch their mental muscles as much as they stretch their legs, wings, and necks.

When Free Range Time Is Over

In a suburban neighborhood, you may have rules against allowing your hens to roam free all the time. When you do allow them to free range, it will probably be when you are in the yard to keep an eye on them. You may find it hard to corral your hens when free-range time is over, and you need to return them to the coop. Take it from a champion chicken chaser, hens can be quick and dodgy. As dusk approaches, though, most chickens will instinctively return to their coop. Let's look at why this is, as well as tips to get your hens to go home.

Chickens Come Home to Roost

Chickens are creatures of habit, and they are driven by strong internal instincts. One of them is the roosting instinct. Chickens know that they are the targets of predators, and they realize that most of these predators come out after dark. So, when the sun starts to set, chickens want to get somewhere safe and high up, like the roosts in their coop. They spend a lot of time in their coop, making it a comfortable and familiar place to be. Out of habit and driven by instinct, most hens will return to their coop on their own at dusk. If they don't, or if you need them to get back in the coop before evening sets, what can you do? Try these tips.

1. Feed Them

Food is a great motivator. Your suburban chickens quickly learn that their food is offered to them in their coop or chicken run. Like Pavlovian dogs, your hens have probably learned a few other cues that indicate they are about to get fed, like the clang of the metal trash-can lid or the rustle of the feed-sack bag or seeing you with a feed scoop in your hand. You can

Photo Courtesy of Madison Woodard

use these to your advantage. When you want your birds to get back in their enclosure, you can, of course, fill up their food dish. Or you can try clanging the trash-can lid or brandishing the feed scoop. That might just do the trick.

2. Stick to a Schedule

Since chickens love routines and habits, you could try putting them on a schedule. Always let them out at the same time and send them back in at the same time. Follow the same routine for both the release and the return. You will be surprised how quickly your suburban chickens catch on.

I have noticed with my hens that they have a free-range schedule that they consistently keep to. They forage around the yard in a pattern and end up at certain points in the yard at the same time each day. It's weird, but I know that if it is three, the hens will be in the southwest corner of the yard. They still don't follow Daylight Saving Time, though.

Say, for example, you always let your hens out in the morning, but they need to coop back up at one so you can leave for work. Stick to this schedule—and maybe bribe them with treats until they catch on—and in a few weeks, you will notice that they habitually head back to the coop around one. Don't call them birdbrains. Chickens are actually pretty smart.

3. Use a Training Whistle

In addition to being able to tell time and pick up subtle clues, chickens are easily trainable. You can teach them to return to the coop on their own by using a training whistle, a song, a light, or any other type of signal. While bribing them with treats or food, blow on your whistle, sing a specific song, flash a light, or do something else that your hens will notice. In just a few days to a week, they will begin to associate that whistle, song, or signal with the treats they are getting in the coop. In no time at all, you will be able to blow the whistle and watch as your hens run back to the coop.

4. Bribe Them

If your hens are not persuaded to return to their coop with a scoop of their layer pellets, try upping the bribe. Mealworms do the trick because chickens absolutely love them, and they have enough of an odor that they will get your birds' attention. Use mealworms or another delicious treat as a reward when you are training your

hens to head back to their home. Treats are a great motivator.

Protecting Free-Ranging Chickens

> *We lost a few hens to hawks. It didn't matter if we were in the yard—the hawks would swoop down and take a hen. We ended up not being able to free range anymore. We came up with the idea of chicken tunnels! This allowed our girls the ability to scratch in the dirt, eat bugs, and roam parts of the yard while still being protected. We quickly realized that it cut down on chicken poop on our deck, and our landscaping also enjoys a break. In the evening, we close up the tunnels so the raccoons can't get in.*
>
> AMY JOHNSON

The safety of your suburban chicken flock should be your number one priority, yet when your hens are free ranging, they are much more vulnerable than they would be in their coop. There are steps you can take to make sure that your hens can enjoy their free-ranging experience while staying safe.

HELPFUL TIP

How Much Space Do They Need?

The USDA's definition of a free-range chicken is any chicken that has "outdoor access." For industrial farm-raised chickens, this outdoor access can be as small as a "pop hole" that doesn't allow for full-body access to the outdoors. Ideally, free-range chickens on a suburban farm should have access to a protected outdoor run to spend hours of their day. When planning a free-range or foraging area for your suburban chickens, you should allot about 8 to 10 square feet of outdoor space per chicken.

1. Have a Fenced-in Yard

Chickens are smart, but they have no concept of property lines. If you don't have a fenced-in yard, you might want to think twice before you let your hens out. First of all, it could be illegal in your area. Some communities have strict ordinances banning free-ranging or requiring specific fencing standards to be met. Second, your neighbors might complain. Even if your neighbors love your hens, they probably won't love chicken poop on their lawns and the mysterious disappearance of all their strawberries. Third, it puts your hens at risk.

Predators can attack at any time, not just after dark. If your yard is not fenced, you are missing the first line of defense against a predatory attack. I have seen a fox come into our yard in the middle of the day. Stray dogs don't wait until night. An unfenced yard poses all sorts of dangers.

It is not just predators either. Without a fence to stop them, your hens could wander into the road and get hit by a car. If you do not have a fenced yard, keep reading. We will talk about an alternative to free-ranging soon.

2. Proper Supervision

Keep an eye on your chickens when they are free-ranging. Stay out in the yard with them so you can make sure they remain safe. It only takes a minute for a predator to attack. If you step inside for a moment, you might be giving a crafty predator the opening it was looking for to spring its attack. Nearly all predators you could encounter in a suburban community will be opportunistic hunters. They are searching for an opportunity to catch a quick dinner, but they aren't looking for a fight. If they see you in the yard, they will move on. Humans are a threat to these animals; they will forgo an easy kill to keep themselves safe.

3. Find an Alternative to Free-Ranging

If your yard, neighborhood, or other situations prevent you from free-ranging your chickens, don't feel bad. You're not a bad chicken parent. In fact, you are a responsible one who cares about the safety of your suburban chickens, respects your neighbors, and abides by the law. You can still allow your hens to explore different areas of the yard, find bugs, and sample new plants by using a chicken tractor or a mobile chicken coop. You can reposition the mobile coop in different spots to give your hens a new view and exposure to different plants, all while keeping them safely confined.

The Drawbacks of Free-Ranging Your Suburban Chickens

> **"**
>
> *Aside from all the obvious considerations like local laws, you need to think about the size of your coop and/or run and where the chickens are going to roam. Will you let them wreak havoc in your flower beds? How will you keep them out of the vegetables? You probably don't want to keep them literally 'cooped up' all the time, but space for a chicken run is space that's not useful for gardening.*
>
> WILLIAM QUIGLEY
>
> **"**

Aside from potential predator attacks and legal restrictions, there are some other drawbacks to free-ranging chickens in a suburban environment. You should carefully consider the risks and drawbacks before you make your decision to free-range or not to free-range.

1. Lawn Chemicals

When you free-range your hens, you could be exposing them to harmful lawn and garden chemicals. Harsh pesticides, herbicides, and fertilizers can sicken your birds or render the eggs inedible. Before you open the chicken coop door, know what chemicals, if any, are on your lawn. It might be safer to keep your hens confined.

2. Chickens Poop

Chickens poop. A lot. You realize this every time you clean out the chicken coop. Just because you're free-ranging your hens doesn't mean you won't have to deal with poop. It just means you'll have to deal with poop in places where you don't want poop to be. An ongoing issue we have is that our free-range chickens like to hop up on our deck. In the summer months, before we enjoy dinner on the deck or invite the neighbors over for a drink, we have to hose down and scrub the deck. Not to be graphic, but the problem is particularly bad when the blackberries are in season, if you know what I mean. If you free-range your chickens, expect poop in your yard and on your sidewalk and, if you are lucky like me, on your deck!

3. Chickens Forage

Chickens enjoy sampling bites out of plants, vegetables, and fruits. When they find something they really like, the whole flock will flock to it to get their share. The cherry tomato plant on the deck was thriving and loaded with almost-ripe tomatoes. And the next day, they were all gone. If you value your landscaping or your vegetable garden, it might be best to keep your hens away from it by keeping them in the coop.

4. Don't Let Your Chickens Cross the Road

Free-ranging chickens may wander into the road. Remember, they need to pick up dirt and pebbles for their digestive system. These are readily available along the side of the road and even on the road. When it comes to a collision between a car and a chicken, the chicken never wins.

Chapter 8 Summary

"Free-ranging" is a popular buzzword among chicken farmers, and with the numerous benefits of the practice, it is easy to see why. However, there are special circumstances regarding free-ranging your backyard chickens that are unique to suburban chicken farmers. Before you let your hens roam free, you need to make sure you aren't breaking any laws or rules. You also need to take steps to keep them from danger when they leave the safety of the coop. Although your chickens will instinctively return to their coop as the sun sets, you can train them to go back into their enclosure on command. Chickens are smarter than you think and will quickly learn their training. If you decide, based on your circumstances, that free-ranging is not for you, you could consider using a chicken tractor to give your hens some of the benefits of free-ranging while they remain safely confined.

CHAPTER 9

Roosters or No Roosters?

> *I would not get a rooster. In an urban setting they are not needed. Roosters are loud! You will upset a neighbor or go crazy yourself listening to it! If I lived in a rural setting and free-ranged my chickens, however, I might get a rooster—they do protect their ladies. Also keep in mind that a rooster doing his 'rooster thing' can leave hens with missing feathers, and sometimes, blood.*
>
> AMY JOHNSON

Roosters are a hot-button topic among suburban chicken farmers. If you are new to backyard chicken raising or if you've never had a rooster in your flock, you might have questions. I must admit, roosters have a bad reputation. Currently, my own flock is rooster-less, but I have had plenty of roosters in the past. I have experienced firsthand the pros and cons of roosters in a suburban flock, which is why I feel it is necessary to devote an entire chapter to the male members of the chicken species.

For many suburban chicken farmers, this is a moot point. They are under local ordinances prohibiting roosters in their area. If you are allowed to keep a rooster, you might be wondering if you should. The information in this chapter will help you understand the challenges of roosters, learn the benefits and drawbacks of including them in your flock, and debunk the myths about roosters. When you balance this information with your personal circumstances, you will be able to decide for yourself whether you want a rooster to move into your backyard coop.

Debunking the Number One Myth About Roosters

When I tell people that I have suburban hens, but I don't have a rooster, without fail, the first question I get is, "Don't you need a rooster if you want to have eggs?" I politely explain that bird biology differs from mammal biology in some significant ways. Of course, my friends are right that a male must be in the picture for a female mammal to have a baby. Likewise, a rooster is required for a hen to have a chick. A rooster is not needed, however, for a hen to lay an egg. Huh? Let me explain.

A hen's reproductive system being what it is, a mature hen will release an egg at the end of each of her reproductive cycles. For chickens, that cycle takes only about 24 hours. If she has mated with a rooster, she will lay fertilized eggs, which can be hatched into chicks. But in the absence of a rooster, the hen will continue to lay her one-a-day egg, but her eggs will be unfertilized. No chick will form in the egg. This might seem weird, but

A ROOSTER AND HIS SPURS

The spurs on the back of a rooster's legs are part of his natural defense system, his weapons to protect himself and his hens from a predator. The spurs are made of keratin and can grow so long and sharp that they can cause some real damage. A rooster's spurs can injure the hens and, if the rooster is aggressive, can hurt you, your family, your neighbors, or the unsuspecting Amazon Prime driver. Removing a rooster's spurs is a controversial subject among rooster owners. How the spurs are removed—well, that's even more controversial.

Removing a rooster's spurs is akin to declawing a cat. While the spurs or claws can be destructive, they are also the only way the animal can protect itself. Without natural weapons, the rooster or the cat is defenseless.

In the past, rural chicken farmers removed the spurs on their roosters themselves, but the technique they used was barbaric and risky. Your grandfather may tell you how he used a steaming hot potato pressed against the rooster's spur to "soften" the keratin, then used a pair of pliers to twist the spur off. It may have gotten the job done, but it was inhumane. The hot potato scalded the rooster's leg. Twisting off the spur was painful to the bird and left a large, open wound that could easily become infected.

Today, this practice is frowned upon. It is preferable to have a veterinarian do this procedure. A veterinarian will typically work on young roosters or cockerels and remove the spurs before they become too long and the core of the spur is not completely developed. After anesthetizing the rooster, the vet will saw off the spur and use a cauterizing tool to keep the spur from regrowing. Afterward, the wound site is sutured closed, and the rooster is given antibiotics to prevent infection.

There is a compromise between the at-home twist-off method and the costly surgical procedure. That is clipping or filing down your rooster's spurs. Since the spur is made of keratin, like human fingernails, it doesn't hurt the rooster to have his spurs clipped or filed, providing you don't take too much off. Like a dog or cat's toenails, there is a vascular core in a rooster's spur. If you clip too low, you will strike the quick, causing bleeding.

To file your rooster's spurs, you can use a sturdy metal file or a Dremel power tool with a sander attachment. If you are opting to use clippers, get a large pet nail clipper like you would use to trim your dog's nails. When you hold your rooster upside down, you put him in a state of tonic immobility and he will no longer flap, fight, or try to resist you. You could also wrap a towel around him to protect his wings. Work quickly to file or clip the spurs and blunt the sharpness of the point. Keep in mind that, like your fingernails, the rooster's spurs will grow back. The rooster pedicure will be a routine maintenance procedure.

it is not so unusual. If you think back to your high school biology days or that seventh-grade human health class, you will recall that female mammals, humans included, release an egg at the end of their reproductive cycle, albeit not a large one with a hard shell.

You will need a rooster if you want to breed your hens, incubate the eggs, and hatch live chicks. However, if you just want eggs to make your morning omelets, your hens can handle that on their own without the assistance of a rooster.

One last thought about this—just because your hens are laying fertilized eggs, that doesn't mean the eggs are inedible. You can certainly continue eating them. If you collect your eggs daily and store them in the refrigerator, you won't have to worry about cracking an egg into your cake mix only to have a baby chick pop out. Hens incubate their eggs for 21 days, and the chicks don't begin to form in a freshly hatched egg for at least the first 24 hours if the conditions are right. The eggs must be kept at a specific temperature and humidity level before the chick can start to develop. In the chilly and dry environment inside your fridge, the chicken development stops.

The Pros of Keeping a Rooster

Despite what you may have heard, roosters have their merits. Some of the roosters I had in the past were wonderful birds and quite handsome too. Are you on the fence about adding a rooster? Here are some of the benefits of keeping a rooster to consider.

Roosters for Breeding

Planning to breed your hens for the chicks? If so, you will need a rooster to mate with your hens so they lay fertilized eggs. Roosters are quite virile. A well-cared-for rooster can service about 10 hens for several years.

Roosters Are Great Watchdogs

Roosters naturally take on the role of protectors of the flock. In the chicken run and when they are free-ranging, the rooster will keep an eye out for danger. If he senses a potential threat, you will hear it, as will the hens. It is not uncommon for a rooster to fight off a predator that is going after a hen. He will fight with his flapping wings, sharp spurs, and strong beak. Many times, the show of aggression by a cocky rooster is enough to send a predator running.

"If it were permitted in my area, I would probably have a rooster," notes William Quigley of Washington. "A rooster stands a chance against a raccoon, the worst predator in a suburban flock."

Amy Johnson of Minnesota agrees. "If I lived in a rural setting and free-ranged my chickens, I would get a rooster. They do protect their ladies."

HELPFUL TIP

Noise Management

Many suburban chicken farmers forgo owning roosters because of the noise they create – but did you know that hens can also make their fair share of noise? Hens typically make two types of noise: scratching/clawing and vocalizations. Vocalization noises are likely to carry beyond the coop's immediate area and are considered "airborne" noises. Various sound-proofing measures can be implemented to minimize disturbances from airborne noises, including insulating your coop or installing thicker siding to muffle the noise.

Roosters Are Nature's Alarm Clocks

The folklore you've always been told about roosters being nature's alarm clock has some truth to it. Many roosters crow at sunup as a way to greet the day and gather their hens. But if you think you can do away with your alarm clock and rely on your rooster to roust you out of bed, think again. Roosters are unreliable. They can't tell time, and they don't care that you have to be at work at 8:00 a.m.

Roosters crow at all times of the day, not just in the morning. And they crow for several different reasons. In addition to announcing the dawn, they will crow to greet you, crow to warn of danger, crow to establish their territory, crow to show dominance over their hens, and crow to hear the sound of their own voices. A crowing rooster adds a charming, provincial vibe to your backyard; however, your neighbors might not feel the same way.

Roosters Are Flock Peacekeepers

Hens, as I have stated earlier, have a strict pecking order and a hierarchy of power among themselves, but their cliques don't include roosters. There are frequent challenges to the established pecking order when a lower-status hen tries to move up the social ladder. When fighting breaks out among the hens, the rooster will mediate the conflict. He knows how to put the hens in their place and stop their petty bullying, thus keeping the peace within the flock.

Roosters Are Quick Learners and Great Teachers

As natural leaders, roosters learn quickly. They will catch on fast when you are training them to go into their coop at night and will, in turn, teach the hens what to do. A rooster takes his job as flock protector seriously and will train his hens to stay near him.

Roosters Are Characters

Roosters are beautiful birds, and each one has its own unique personality. They add spunk and flair to your flock. If you think hens are comical to watch, try observing a rooster with all his boastful strutting and cocky crowing. Although some have stronger personalities than others, they all add a new level of entertainment to your backyard antics.

One of my favorite roosters was Rocko, an Old English bantam rooster. He was rather handsome, with an auburn back and head and an elegant spray of black tail feathers. My daughter showed Rocko at some local poultry shows, and he relished the attention. He was the big man in the chicken yard. The whole family enjoyed our interactions with Rocko, who behaved like a proud, well-dressed dandy, especially when he was trying to impress the ladies.

The Cons of Keeping a Rooster

Having a rooster around isn't all fun and games, though. There are good reasons, as we will see, why many suburban communities have a "no rooster" clause in their backyard chicken ordinances. Let's look at the drawbacks of keeping a rooster in your flock.

Roosters Will Annoy Your Neighbors with Their Crowing

Crowing comes naturally to roosters. They don't care if it is the crack of dawn or when your neighbor's baby is trying to nap, or when your second-shift neighbor just got home from a long evening at work. They will crow when your neighbor is in the middle of an important Zoom meeting or during a backyard birthday party. None of these will endear you to your neighbors. You can expect some irate phone calls and angry knocks at the door. At best, you will get hostile glares.

As Amy Johnson explains, "Roosters are LOUD! You will upset a neighbor or go crazy listening to it yourself."

You cannot train your rooster to keep quiet or to only crow at certain times. It is easy to see why the noise associated with a rooster is the number one reason why many neighborhoods prohibit roosters for suburban chicken flocks.

Roosters Can Be Aggressive

We once had an aggressive rooster. We named him "Sunday Dinner" because we repeatedly told him, "If you act up again, you're going to be Sunday dinner!" Our threats meant nothing to him. Our young children were the targets of his unprovoked attacks. Sunday Dinner leaped at them, madly flapping his wings and lash-

ing out with his long, sharp spurs. It got to the point that the kids were afraid to go out to do chicken chores by themselves. No, Sunday Dinner never ended up as a meal at our house. Another family from our 4-H club offered to take him to their large country farm, and we were happy to see him go.

Not every rooster is aggressive, but the ones that are can be downright mean. It is instinctive for them to attack anything that might threaten the hens, but sometimes, as in the case of Sunday Dinner, their threat radar is out of whack. If you opt to include a rooster in your flock, spend plenty of time socializing him when he is a chick and cockerel so he is used to everyone in your family. He may not view your family members as threats if he is familiar with them. This is not a foolproof method, however. Some breeds of roosters are simply more aggressive than others.

Your safety and the safety of your family should be your top priority. An aggressive rooster might be an unnecessary danger that is not worth the risk.

Roosters Can Mistreat the Hens

If you notice that the feathers on your hens' backs are broken or missing, it could be that your rooster is being too rough with them during the mating process. When the rooster mounts a hen, he holds the back of her head with his beak and balances on her back with his claws.

His movements, if they are too harsh, will cause the hen's feathers to break off, leaving patches of bare skin.

"Keep in mind," notes Amy Johnson of Minnesota, "a rooster doing his 'rooster thing' can leave hens with missing feathers and, sometimes, blood."

This can be more of an issue in a suburban flock with a limited number of hens. Instead of spreading his love around to several hens, the rooster only has a few of them at his disposal. The more frequently he mates with the hens and the rougher he is on them, the more the hens will suffer.

More Than One Rooster Spells Trouble

In a suburban chicken flock with a limited number of hens, more than one rooster will cause no end of trouble. Like hens, multiple roosters will

establish a hierarchy of power, with one bird as the dominant rooster and the others as his sidekicks. One rooster can adequately handle about 10 hens by himself. And he doesn't like to share. If you have two roosters and less than 20 hens, the roosters will fight over the hens. If their enclosure is not roomy enough, that will cause them to bicker as well. Their fights will be violent and bloody and traumatic.

Maintaining a good rooster-to-hen ratio is the first step to preventing rooster fights. Most likely, though, you will not be permitted to keep the number of hens needed to keep two or more roosters happy. The best solution is to keep only one rooster or to not have a rooster at all.

The Legalities of Keeping a Rooster

You must check the laws and ordinances in your community before you welcome a rooster to your backyard chicken coop. You don't want to face fines or legal repercussions for violating an ordinance. Contact your town hall or city zoning office to learn what the regulations are regarding chickens in your area. An official at either of these offices will be able to tell you the number of chickens you can have and whether or not a rooster can be included in that count.

Also, contact your neighborhood homeowners' association and find out their rules and restrictions. You may live in a town that permits roosters, but if they are prohibited by your homeowners' association, you will have to reconsider.

Even if you learn that you can legally have a rooster in your suburban flock, it would be the neighborly thing to discuss this with the people who live closest to you. You don't want to upset your neighbors, which will disrupt the harmony of your home, by springing a rooster on them. Craig Hansen of Iowa agrees. "I would discuss it with your neighbors before bringing a rooster home."

Yes, it is your backyard, and yes, you can do what you want, but be respectful, understanding, and polite. Give your neighbors a realistic idea of what to expect from the rooster and his behavior. Ask them if they have any concerns or questions. Let them know they can reach out to you if they experience any problems with your rooster. You may want

to smooth things over with them by dropping off a carton of eggs every so often. Keep the lines of communication open and remain respectful and understanding of their point of view. That won't keep your rooster from crowing at inappropriate times, but it will help you keep peace in the neighborhood.

If you think that local ordinances or your homeowners' association are unfairly preventing you from owning a rooster and you believe you can make a strong case for keeping a rooster, you can seek legal advice. An attorney might be able to help you plead your case.

Tips For Keeping Roosters in a Suburban Setting

After weighing the pros and cons, you may decide that having a rooster as part of your suburban flock is a good idea. Before you jump on the rooster bandwagon, you should take steps to minimize the drawbacks. How can you do that? You could try these tips.

Give Your Rooster Plenty of Room

Some of the aggression, noise, and rough treatment of the hens can be minimized by giving your rooster plenty of room. When the coop gets too crowded, your rooster might be so frustrated and irritable that he takes things out on you, your family, and the hens.

Unless you have a large flock of hens, keep only one rooster. If your rooster-to-hen ratio is askew and you have more roosters than you need, the roosters will fight it out for top access to the hens. It will get ugly. Fighting roosters can seriously hurt each other, or worse. They can also harm your hens, your family, and you.

Have a Rooster Jail

To minimize the harsh treatment of the hens, prevent roosters from fighting, and keep you and your family safe, consider adding a rooster jail to your coop. Like a mini coop, a rooster jail is a secure, enclosed housing unit you can use if and when it is necessary to separate your rooster from the rest of the flock, even for short periods of time. The rooster jail, like a chicken coop, should be secure, well-ventilated, and protect the inmate from the elements.

A warning, though, rooster jails occasionally backfire. Instead of using solitary confinement to cool their anger and think about what they've done, some roosters become more frustrated and aggressive from being pent up.

Train Your Rooster

I have never had much luck training a rooster to be kinder and gentler, but I have been able to use his coop training to distract him from time to time. If you trained your chickens to respond to a whistle, a song, or a command word when you want them to return to the coop, you can give this command to send your rooster back to his enclosure if he is acting aggressive or bullying the hens. Roosters are quick learners and will associate the command with getting a treat; therefore, they will respond to the command faster than the hens. It'll give you a chance to lock the rooster up and keep him separated from the flock.

Only Keep One Rooster

With another rooster on the scene, your rooster will become more aggressive. Both birds will try to one-up each other, causing sparring and bickering. It gets worse when they start fighting over access to the hens, especially if there aren't enough hens to satisfy both roosters. For a small, suburban chicken flock, one rooster is more than enough.

Seek Advice from Experts

Talk to experienced chicken farmers, experts at your local extension office, or even a veterinarian. They may have tips or suggestions to help you manage your rooster. For us, a valuable source of advice was our 4-H club. There were several families in our club that had raised chickens for years. Any situation or problem we encountered, they had dealt with it in the past and could tell us how best to handle it.

The folks at your county extension office might be able to put you in touch with other chicken farmers in your area or direct you toward other resources. You could also contact your veterinarian. One discussion you could have with your veterinarian is about removing your rooster's spurs. Take a look at the sidebar for more information about this.

Get Rid of Unwanted Roosters

When you purchase straight-run chicks, you are getting hatchlings that have not been separated by gender. In theory, a straight-run batch should contain roughly half males and half females, but that's usually not the case. At the hatchery, workers pull out the females to fill orders for pullets. That skews the ratio, so straight-run orders tend to have more cockerels than pullets. What that means is that you will end up with way more roosters than you need. How do you get rid of the excess cockerels?

You have a few choices, and some are not pleasant. The young cockerels could be culled or raised until they are big enough to be butchered for their meat. Or you could reach out to rural chicken farmers in your area to see if anyone could rehome your roosters. Kristen Harrell of Iowa explains, "I have given up all my chicks that were roosters. In my area, a rooster is not allowed."

Chapter 9 Summary

Despite the negative press they may have gotten, roosters can be a welcome addition to a suburban chicken flock, that is if you are legally allowed to keep a rooster in your community. A rooster can protect your hens, add personality to your flock, and serve as the peacekeeper between the hens. And of course, if you intend to breed your hens, you will need a rooster. However, roosters are loud. A rooster can be rough on the hens or hostile to you and your family.

So, how do the pros and cons of keeping a rooster add up? That is entirely up to you. For suburban chicken farmers who live where roosters are permitted, it is a personal choice whether you decide to include a rooster or keep your flock a girls-only club.

CHAPTER 10

Wintering Over with Chickens

Y ou may have welcomed your chicks in the early spring and watched them grow and thrive all summer long. Your suburban flock has settled into your backyard coop, started supplying you with eggs, and enjoys foraging in your gardens. All is well—until the first snowflakes start to fall. Now what? Can your hens survive the winter in their henhouse? The answer is yes, they can.

"Chickens are built for the cold weather," explains Elizabeth Sorby of Washington, "as long as they can get out of the wind into an enclosed coop. They huddle together and fluff up their feathers—think of down comforters—to keep warm."

Chickens are quite hardy, but wintertime poses some challenges to suburban chicken farmers. In this chapter, we will offer tips and advice for wintering over your chickens.

Winter Is Coming

> "
>
> *Chickens are built for the cold weather as long as they can get out of the wind into an enclosed coop. They huddle together and fluff up their feathers (think of down comforters) to keep warm. I've never used heat lamps in their coops in the past 15 years of keeping chickens. When it is sub-freezing, you do need to warm their water or change it frequently throughout the day.*
>
> ELIZABETH SORBY
>
> "

Rather than ignore the calendar and deny the increasingly early sunsets, you should make some preparations to protect your flock from the freezing temperatures and snow. In the months leading up to winter, you can take steps to store chicken feed in cold weather, winterize your coop, stop drafts, and provide additional lighting.

FUN FACT

A Distinguishing Palet

While chickens are known for being indiscriminate consumers of insects and produce, these feathered friends can actually distinguish between several flavors. This sense of taste is due to the chickens' average of 240-360 tastebuds.

Storing Food in the Winter

Chicken feed, particularly layer pellets, contains moisture. When the temperatures dip below zero, the feed pellets can freeze and fuse together. Frozen chicken feed is almost impossible to scoop out. You have to practically chisel it loose. Over the winter months, consider storing your feed in an insulated storage container. I have a friend who uses an insulated cooler with a latch closure to store chicken feed in the winter. You could also move your feed container into your garage, porch, or mudroom to prevent the feed from freezing.

The feed will also freeze into a solid lump in your chicken's feed dish, but let's talk about that problem in a moment.

Insulate Your Coop

In the warm and humid summer months, it is important for your backyard chicken coop to be open, airy, and well-ventilated. Chickens suffer more in the heat than they do in the cold, but that doesn't mean your suburban flock will be comfortable in their airy summer coop when temperatures dip low.

Consider adding insulation material to your coop to keep out the cold. If the interior of the henhouse has studded walls, you can install rolled fiberglass insulation or insulating fabric between the studs. Then cover the insulation material with plywood or thin particle boards. Don't leave the insulating material exposed, or your chickens will peck at it, especially when they are bored. The materials can be harmful to the hens. If you are using a wool insulating fabric, you can tack it up on the exterior of the coop.

Protecting Against Drafts

Although good ventilation in your suburban chicken coop is just as important in the winter as it is in the summer, too much airflow can be a problem. The chickens' body heat won't remain inside the coop if it is too drafty.

Look for cracks and holes in your coop that you can fill with caulk. You can insulate your coop by putting plexiglass over screened windows or openings. If your coop is located in a spot that catches the wind, you could tack a plastic tarp on the windward side to block the wind but don't encase the entire coop in plastic wrapping. The chickens still need good airflow. This prevents moisture buildup that can freeze.

Make a Windbreak

You can take a step toward reducing drafts in your suburban chicken coop by creating a windbreak. If you have a mobile chicken coop, you can move it to a sheltered spot in your yard, like near a privacy fence or a hedge row. This will block some of the wind from hitting the coop. You can also try securing a plastic tarp to the fencing on the side of the chicken run. Avoid extending the tarp to the top of the fenced enclosure. We did this once, thinking we were being smart and proactive. The first snow of the year was a wet, heavy snow that collected on the tarp. The weight caused the entire chicken run to collapse. Oops. Fortunately, the chickens were all safely in the coop when this happened.

Use Extra Bedding

During the winter months, increase the amount of bedding you use in your backyard coop. Chicken farmers call this the deep litter method. A good layer of shavings, straw, or similar organic material covering the floor of the coop provides extra insulation and gives your hens a place to snuggle down and keep warm. Replace the bedding often—it will become soiled faster because the hens are spending more time inside—to keep the bedding material from becoming compressed. It should be loose and light so the hens can burrow down in it.

Install a Light

While you are making upgrades to your backyard chicken coop to winterize it, you might think about installing a light. There are two reasons for this—one for you and one for the birds.

When the number of daylight hours diminishes in the winter, you might find yourself doing chicken chores in the dark. A light will make this task easier and safer for you.

As we will discuss in more detail in the next chapter, light is a factor in egg production. Most suburban hens stop laying during the winter months, not because of the cold temperatures, but because they are not receiving the appropriate amount of sunlight needed for them to produce eggs. You can supplement natural sunlight with an electric light and hope to keep egg production going all winter long.

Use the Sun's Heat

A benefit of covering chicken coop windows with clear plexiglass sheets is that it lets the sunlight in. Hens, as I mentioned, need to be exposed to sunlight to keep their egg production from dropping off. The plexiglass has an added benefit. It creates a greenhouse effect in your coop. The sun's rays penetrate through the clear plexiglass and warm

up the coop, but the heat cannot escape. On a sunny January day, the temperature inside the coop can rise several degrees.

Provide a Heated Water Dish

Frozen chicken waterers are the absolute worst part of winter. My life turns into a cycle of thawing frozen water dishes and replacing waterers several times a day. As glamorous as that sounds, you can avoid my struggles by using a heated water dish or chicken waterer. Like heated dog water bowls, these heated waterers have a long, insulated electric cord and heat coils that prevent freezing. Of course, to use these, you need to have a nearby electrical source, which is my problem.

DON'T Use a Heat Lamp

It is tempting to want to keep your hens warm by taking the heat lamp from your brooder box and hanging it in your backyard coop. Nearly every chicken expert you encounter will tell you the same thing. The risk of fire is just too great. Heat lamps put off a lot of heat, and your chicken coop is packed with flammable materials, like wood shavings or straw.

I am embarrassed to share this story, but if it helps just one suburban chicken farmer keep his or her flock safe, then it is worth my red face. Years ago, early on in our backyard chicken journey, we hung a heat lamp in our coop to keep the hens warm on a frigid night. In the wee hours of the morning, my volunteer firefighter husband's pager started going off. His department was being called out because "a passerby sees flames coming from an outbuilding." And then, the 911 dispatcher gave the address. Yep, OUR address! We ran outside, in pajamas and barefoot in the snow, but it was too late to save any of the six hens inside. In fact, the coop was nothing more than a pile of smoldering ashes by the time the fire engines rolled up. It doesn't take long for a small wooden coop filled with wood shavings to go up in smoke.

Of course, we had the heat lamp securely hung high off the ground, well away from the bedding. We can only speculate, but we think one of the hens either tried to roost on top of the heat lamp or flew up to it, somehow knocking it down.

My whole family was devastated by this fire, and the kids still speak fondly of the hens we lost that night. I feel it is an important story to tell. I hope other suburban chicken farmers learn from our tragedy.

Admittedly, we were all really stupid. As a trained firefighter, my husband should have known better, and, of course, he endured weeks of razzing by his fellow firefighters. The razzing from the chicken coop fire, however, was mild compared to the teasing he got after one of our goats fell into an old well, and we had to call the fire department to get her out! But that's a story for another day.

Don't take my word for it. Here is what our panel of experts had to say about heat lamps.

"I don't ever recommend using heat lamps in a coop because I have seen too many stories about fires caused by heat lamps," explains Craig Hansen of Iowa.

Elizabeth Sorby of Washington adds, "I've never used heat lamps in my coops in the past 15 years of keeping chickens."

Caring For Your Chickens in the Winter

> *Frozen water is always a concern. I have used heated dog bowls and heated waterers with good results. I don't ever recommend using heat lamps in a coop because I have seen too many stories about fires caused by heat lamps.*
>
> CRAIG HANSEN

During the cold winter months, your chickens need some extra care to keep them healthy and happy. The freezing temperatures naturally create obstacles to you as a suburban chicken farmer. Maintain a close eye on your flock, maintain a consistent feeding and watering routine, and try to keep your hens entertained.

Maintain Access to Food and Water

It is interesting to note that frozen water is the number one winter-time challenge, as noted by our panel of experts. It's good to know that I'm not the only one dealing with solidly frozen waterers all winter long. According to Amy Johnson of Minnesota, the biggest challenge is "keeping the water from freezing! A heated waterer is a must."

"When it is sub-freezing," says Elizabeth Sorby of Washington, "you do need to warm their water or change it frequently throughout the day." Craig Hansen of Iowa adds, "Frozen water is always a concern. I have used heated dog bowls and heated waterers with good results."

In the winter months, you will need to visit your backyard chicken coop more frequently to make sure your hens have access to food and water. Yes, this means trudging through knee-deep snow and battling below-zero windchill several times a day, but you shouldn't need to stay out there too long.

If the layer pellets are frozen together into a solid block in the chickens' food dish, you will need to break it apart or replace it with fresh pellets. Remove the frozen waterer, unless you have a heated one, and replace it with another waterer filled with fresh, unfrozen water. Even if you are using a heated water dish, be sure to check it often and refill it as necessary. Chickens need more food and water during cold weather.

Replace or Refresh the Bedding

Since your hens are spending more time in their coop in foul weather, they will soil the bedding material quicker than they did in the summertime. As often as necessary, rake out the old bedding and replace it with an extra thick layer of new, clean bedding. Chickens need more bedding in cold temperatures to help them keep warm. If your bedding material is still fresh enough but appears to be settling down, you can loosen it with

a pitchfork or add another layer of bedding on top of it so the hens have a clean, dry place to snuggle.

"One of the biggest challenges is keeping the coop clean because the chickens are not in the yard as much, and the poop collects much quicker," says Kristen Harrell of Iowa.

Monitor the Health of Your Hens

Photo Courtesy of Madison Woodard

On your frequent visits to your backyard chicken coop, take a few moments to look over each of the chickens to make sure they appear healthy and uninjured. See if any of the birds look weak or lethargic. Look for any open wounds or signs of bleeding.

Remember that bored chickens might pick on each other. Or worse, a group of hens might target a weaker chicken. They might pull out the vent feathers on their poor victim, peck at her comb, or prevent her from eating. That leads me to the next topic.

Provide Stimulation and Entertainment

You know how bored you get when you are cooped up in the house all winter. Your chickens feel the same way, only they can't entertain themselves by streaming the latest hit series or catching up on their reading. It is up to you to provide some sources of stimulation and entertainment for your hens. Here are a few ideas.

Buy a small pumpkin and drill two holes in the sides. Thread a piece of twine or a small chain through the holes and suspend the pumpkin from the ceiling of the coop. Make sure it hangs low enough for the hens

Photo Courtesy of
Anne Greenwood

to reach it. They will spend hours pecking at it. Every peck will send the pumpkin swaying or spinning, which the chickens will find entertaining.

Instead of a pumpkin, you can string together a few apples. Chickens are attracted to the red color and, like the swinging pumpkin, the challenge of pecking at a swaying apple will occupy the hens for hours. Switch it up and hang a head of cabbage. Chickens love that as well.

Stop by your local pet store and purchase some live crickets. These are sold as food for lizards and snakes, but chickens love them too. And crickets are a great source of protein. Let the crickets loose in the hen-house, and your chickens will run around hunting them down. You might want to save this activity for a milder day. Frozen crickets aren't as much fun to catch.

You can find recipes online to show you how to make a "flock block." This is basically melted suet that is mixed with scratch, whole corn, meal-worms, and other treats. The mixture is poured into a loaf pan or other type of mold. When it is cool and hard, you can hang it in your chicken coop for the hens to peck at.

There are several kinds of puzzle feeders for chickens available for purchase online, or you can make your own. Just take an empty plastic water bottle and drill several holes in it. Make the holes about a half-inch in diameter. Fill it with scratch or layer mash—maybe toss in some mealworms to liven it up—and roll it toward your hens. As it rolls, food will fall out of the holes. Chickens are smart. They will quickly discover that the more they roll it, the more food spills out. Soon, they will be expert soccer stars!

Let Them Fly the Coop

There is no need to keep your suburban chickens locked in their coop all winter long. Even if there is snow on the ground and a nip in the air, you can still let your hens outside. They will welcome the opportunity to stretch their legs, shake the dust off, and explore the snowy backyard. If it is a fairly mild winter day and they aren't out for hours on end, you don't need to worry about them getting chilly. After all, they are like Elsa from Disney's *Frozen*. The cold never bothers them anyway!

Tips for Keeping Your Chickens Healthy Throughout the Winter

> **"**
>
> *Fun fact: Chickens can actually eat snow as a form of water. I prefer a heated water dish. I also wrap my chicken run in clear plastic tarps to keep the snow out. My urban chickens don't enjoy standing on snow. A cozy coop heater works great, and my hens snuggle up at night. Having an electronic thermostat/hygrometer is really useful. I can monitor the temp and humidity from inside my house. I go all-out during the winter and have lights on timers, heat, and a heated egg box!*
>
> AMY JOHNSON
>
> **"**

We've mentioned your chickens' health several times in this chapter, but it is worth mentioning it again. The cold temperatures in the winter months pose a new set of challenges to keeping your hens in good health. Follow these tips to make sure your hens remain in good health when temperatures dip low.

1. Provide High-Energy, High-Protein Treats

Chickens expend a lot of energy keeping warm. You can give them all the energy they need by offering high-energy treats. Nuts and seeds are a great choice. Try pumpkin seeds, black oil sunflower seeds, shelled walnuts, almonds, or shelled peanuts. Another great source of protein for chickens is sprouting beans. Lentils and peas, too. Suet and tallow also provide extra protein. Of course, chickens love mealworms.

2. Protect Against Frostbite

Hens don't have combs and wattles that are as large as those on roosters, but these body parts are quite vulnerable to frostbite. Look for black or blue spots on the combs and wattles, blisters, bleeding, and loss of tissue structure. As you can imagine, this is very painful. You can easily protect your chickens' wattles and combs in extreme weather. Just coat the combs and wattles with petroleum jelly to insulate these exposed features from the bitter cold and the potential of frostbite.

3. Give Them a Hot Meal

Some heated mash, poultry porridge, or warm oatmeal will warm your hens up from the inside out. The hens will appreciate the kind gesture. You can mix layer mash with warm water or use hot, homemade bone broth. Most store-bought broth contains too much salt and is not good for your chickens, so stick to making your own broth. Broth, especially bone broth, is high in protein and calcium—two things your hens need in the wintertime.

Egg Production in the Winter

We will go into more detail about egg production and the science behind it in the next chapter, but you should be aware that egg production drastically slows down or halts altogether in the winter months. The reason for this, as we will see later, is linked to sunlight hours, not temperature.

The decline in egg laying in the winter is part of chickens' biological adaptations. It is possible, however, to trick the hens into continuing their egg production by installing electric lights that mimic natural sunlight.

Most suburban chicken farmers simply allow their hens to slack off in the winter. If your hens continue to lay eggs throughout the winter months—either because you are using an artificial light source or because they are continuing to lay on their own—you will need to collect eggs more often to prevent them from freezing. When eggs freeze, the shells crack, allowing bacteria to get in. You will need to discard them. This waste can be avoided by keeping the nesting boxes well-filled with straw or other bedding material and by collecting eggs before they have a chance to freeze.

Chapter 10 Summary

Wintertime creates more challenges for suburban chicken farmers, but with some advanced preparation and some extra TLC, your hens will weather the cold and be ready to welcome spring's warmer temperatures. Be sure to inspect your backyard coop, make any necessary repairs, and winterize it. You can keep your chickens active and entertained by offering treats in creative ways, like suspending a pumpkin or head of cabbage for them to play with.

Be sure to allow your chickens to go outside on milder days. They will welcome the opportunity to leave the coop and get some fresh air. Keep in mind that hens may stop laying eggs over the winter, but if you are still getting eggs, you will need to gather them more often.

In the cold temperatures, you can prevent painful frostbite by applying petroleum jelly to your chickens' combs and wattles. And you can spoil them with hot food, like heated mash. Winter might present some obstacles for suburban chicken farmers, but chickens are hearty and resilient.

CHAPTER 11

Egg Production

Eggs. That's the number one reason why suburban residents want to raise chickens in their backyard in the first place. Chickens are amazing, egg-producing machines—as long as the conditions are right. To know what those ideal conditions are, we first need to understand how eggs are formed. In this chapter, let's discuss how chickens form and lay eggs, the factors that contribute to efficient egg production, and how to help your suburban hens be more productive. We will also offer tips for collecting and storing eggs. Lastly, let's discuss the issue of a broody hen.

The Egg-Laying Cycle

A hen's egg is the bird's reproductive cell. Within the hard shell, there is a yellow yolk and a white liquid albumen, which will magically become a baby chick if the egg has been fertilized. Yolk material forms in an organ called the follicular sac. In fact, in mature hens, this sac continuously creates the yolk material. The yolk of a specific egg is made about seven to nine days before the hen lays the egg. When the follicle sac releases the yolk, it forms a thin membrane around it that holds the yolk together. Ovulation occurs when the yolk is released from the follicle sac.

The released yolk moves from the follicle sac to the oviduct in the hen's abdomen. Here, the albumen, or the thick white portion surrounding the yolk, is added. Another thin membrane holds all this together. At this point, the yolk and albumen move into the magnum, where the first shell layer is formed around the yolk.

Next, the egg progresses through to the hen's uterus, also referred to as the shell gland. The egg remains in the hen's uterus for about 20

hours as the hard calcium carbonate shell is added. Lastly, the egg passes through the hen's vaginal canal. Here, a thin coating of a protein substance is applied to the shell. This is called the bloom. The egg's shell is porous, so the job of the bloom is to prevent dust, dirt, and bacteria from seeping into the egg through the shell.

The hen lays her egg, large end first, from her vent. The entire process of egg formation takes between 24 and 26 hours, but the egg making is a lot like an assembly line. When one egg is in the vaginal canal getting its final touches, there are new eggs forming in the follicle sac, oviduct, magnum, and shell gland.

Unlike human females and other mammals, hens have just one ovary and oviduct; however, it is a highly efficient system that produces eggs on a daily or almost daily basis. Chickens living in a suburban setting can live up to eight years and will be productive egg layers for four or five years. If you run the numbers, that's a whole lot of eggs!

When Will My Chickens Start Laying Eggs?

In general, you can expect your pullets to lay their first eggs when they are between 18 and 22 weeks old. Some breeds of chickens, like Rhode Island Reds and Leghorns, typically mature a bit earlier than heavier-breed hens, like Barred Rocks and Orpingtons.

Age and breed, however, are not the only factors dictating when your hens will begin laying. Nutrition and sunlight are also vitally important.

Nutrition

Chickens have a natural biological adaptation that puts a pause on the reproduction cycle if the hen is malnourished or underweight. It is a survival instinct that makes sure new chicks are not hatched into a situation in which there are not enough food resources for them to survive. In a suburban setting, where the flock is well cared for, this is usually not a factor.

Sunlight

The process of egg production in chickens is activated by photostimulation. The pineal gland in the brain of the chicken processes incoming sunlight, which triggers the release of serotonin and melatonin. These hormones regulate the hens' circadian rhythms. The optimal amount of light per day to stimulate egg production is between 14 and 16 hours.

Photo Courtesy of Kelly Kneeland

This is why chickens will lay more eggs in the spring and summer and why egg production drops off in the late fall and winter.

Commercial poultry farms use artificial lighting to keep hens laying year-round. It is important to note, however, that the type of light is just as important as the amount of light. Remember back to middle school science class when you learned about ROYGBIV and the light spectrum—the different-colored wavelengths that make up the visible light spectrum? Red, orange, yellow, green, blue, indigo, and violet? The type of light wavelengths that hens need for egg production is the higher wavelengths in the red and orange part of the spectrum. These wavelengths are better able to be processed in the hen's brain to stimulate egg laying. If you decide to use artificial lighting in the fall and winter, look for bulbs that are specially made for this purpose, or look for bulbs that mimic natural sunlight.

DID YOU KNOW?
Annual Egg Count

In optimal conditions, a well-fed suburban chicken can produce up to 250 eggs in their first year. After that, a chicken's egg production will naturally diminish as she ages, with a sharp dropoff at around six or seven years old. Most chickens do not start laying until they are 18 weeks old, after which they can produce an egg daily. The Guinness Book of Records recorded the highest number of eggs laid by a single chicken, 371 eggs in 364 days.

Factors that Impact Egg Production

In addition to sunlight and proper nutrition, stress, dehydration, weather, health, and molting can play a role in egg production in mature hens.

Stress

In the bravery category, chickens are—well, chicken! They scare easily and get upset when things change in their world. When they experience a sudden or intense fright, the stress of it will cause a hiccup in their egg production. Last summer, a crop duster sprayed the farm field behind our house. The loud, low-flying, bright yellow plane totally freaked out my hens. They didn't lay eggs for three days! Chickens can get stressed out by the neighbor's new dog barking all day or the road crew repaving your street. A predator scare will also cause your hens to stop laying for a day or two. If a hawk has attempted an aerial attack or a raccoon tried

to get into the henhouse, your chickens might be so rattled that they put egg production on the back burner while they calm their nerves.

Lack of Water

Chickens need to have constant access to clean, fresh drinking water. If that access is interrupted—by a frozen water dish or if the waterer gets knocked over—and the hens become dehydrated, they will not lay eggs. Portions of the egg, like the albumen and the yolk, are composed of water. If the bird is dehydrated, its body pulls hydration from the reproductive system to other vital bodily functions.

Weather

Extreme heat or extremely cold temperatures will disrupt egg laying. Humidity levels that are too high or too low can also contribute to a slowdown in egg production. Ideally, hens prefer temperatures between

55°F and 80°F, with humidity levels between 40 to 60%. That doesn't mean they will stop laying if those conditions are not met; it just means they may not be at peak efficiency at that time. Likewise, a severe thunderstorm with lightning and high winds might stress your hens so much that they refuse to lay for a few days.

Illness, Infection, and Infestation

Hens that are fighting off an illness, infection, or infestation will temporarily halt egg-laying. Their bodies are too busy dealing with health problems to devote reserves to reproduction. Chickens are susceptible to contagious diseases, especially respiratory illnesses, but parasites, like lice, mites, and worms, can also decrease egg laying. If a hen has been injured and is recovering from that, she may not lay eggs until she is fully recovered. Her immune system is working extra hard to keep infection away.

Molting

All birds molt. This is the natural cycle of dropping old feathers and regrowing new ones. Chickens need extra protein during the regrowth stage to produce their new feathers. If hens do not have a high level of protein in their diet, they won't have enough supplies to grow feathers and create eggs at the same time. You can boost your hens' protein intake during their molting period to increase the probability that the hens will continue laying through their molt.

Photo Courtesy of Brian and Jen Jacob

Tips for Improving Egg Production in the Suburban Setting

There are many parts of the egg-laying process that are beyond your control as a suburban chicken farmer. However, you can take steps to create an ideal egg-laying environment for your hens, which will improve their egg production. Here are a few tips.

1. Select the Right Breed of Chicken

If your goal for your suburban flock is to have optimal egg production, your first step is to make sure you buy the right breed of chicken. Chicken breeds are categorized as layers, meat birds, or dual-purpose chickens because these specific breeds have been developed for these purposes. In a suburban setting, you probably aren't allowed to butcher your own chickens for the meat; therefore, the top reason for keeping chickens is for their eggs. Knowing that, you should focus on layer breeds for your backyard flock.

Some of the best layer breeds for optimal egg production include Leghorns, Plymouth Rocks, Rhode Island Reds, Australorps, and Buff Orpingtons. These breeds not only lay more eggs, but they typically start laying at an early age and continue laying longer in life than other breeds.

2. Provide Adequate Housing

Happy hens will be the most productive. By providing a safe, secure chicken coop and chicken run, you are taking a big step toward having a happy flock. Your hens will be less stressed when they are in a safe environment that keeps them sheltered from the elements. They also need a home that is clean, well-ventilated, and roomy enough for all the chickens. If their home life is not ideal and is causing them stress, it will manifest in fewer eggs.

3. Offer Nutritious Food

Hens need proper levels of vitamins, minerals, protein, and calcium in their diet to keep their egg-laying skills in tip-top shape. By feeding them a good-quality chicken feed that is formulated for layers, you will be giving them the fuel they need. Supplementing their layer mash or layer pellets with high-protein mealworms, calcium-rich oyster shells, and grit to aid in digestion will give them an added boost of nutrients. Also, be sure their waterer is always full so they have continual access to fresh water.

4. Give Them Plenty of Light

Now that you understand how important sunlight is to egg production, you can make sure your hens get as much light as possible. This means the henhouse should have windows or a skylight to let the sunshine in. The hens should have a secure chicken yard or chicken run so they can safely spend time outside, soaking in the sun.

In the late fall and winter, when the days are shorter, you can install supplemental lighting if you hope to encourage your hens to keep laying. Just make sure you are using the right bulbs and that you are following good safety practices to prevent a fire.

5. Give Them a Chance to Exercise

Exercise and activity will help your hens stay healthy and fit, which will, in turn, improve their egg production. Allowing your suburban chickens to spend time in their chicken yard and forage in your yard will get them moving. I purposely place my hens' feed dish and water dish at opposite corners of their enclosure. This forces my birds to walk more. When you use stimulating activities, like a puzzle feeder or a hanging head of cabbage, your chickens will get some extra exercise while they stimulate their brains.

Collecting and Storing Eggs

This is the exciting part. After caring for your hens from the time they were tiny chicks, you are now rewarded with a supply of fresh eggs. There are a few things to keep in mind when it comes time to collect eggs.

Don't Wash Eggs

I know this goes against your instinctual habits, but don't wash the eggs after you collect them. When you wash the eggs, you are actually increasing the chances of contaminating the eggs. It doesn't sound logical, but hear me out.

Photo Courtesy of Lana Plashchynskaya

Remember a page or two ago when we explained that the last step in the egg-making process, just before the hen lays the egg, is when the eggshell is coated with a protein substance called the bloom? The bloom is a microscopically thin layer of material on the outside of the eggshell that prevents bacteria, dirt, and dust from getting through the tiny, unseen pores in the eggshell. This is nature's way of sealing the egg and protecting its contents. When you wash your eggs, you remove the bloom and open up the eggs to contamination.

Instead of washing your eggs after you collect them, gently wipe off any dirt or chicken poop that might be on the eggs. Then trust nature. The egg's bloom will keep its contents contamination free.

Acquire Egg Cartons

You will need some egg cartons to store your eggs. I have always had good luck just asking my friends, family, and coworkers to save their cartons for me. We have also purchased cartons, either online or at a local feed store. When you buy them in bulk, they are fairly cheap—less than a dollar per carton. One benefit of buying egg cartons is that they are blank, with no printing on them. For a while, we had our own cute and clever labels to affix on the cartons when we gave them to friends or family.

Collect Eggs Often

Once your hens start laying, you should make a habit of collecting eggs a few times a day. Your goal should be to remove the eggs from the nesting boxes as soon as possible so that hens don't have a chance to crack the egg or soil it or peck a hole in it. Many suburban chicken farmers will suggest that you collect the eggs in the morning and again in the evening. I would suggest that you gauge your own flock to see if there is a consistent pattern to their timing. I have learned that my hens lay mostly in the morning. It is rare for one of the hens to lay an egg after noon. Your hens, however, may have a different schedule.

In very hot weather or freezing weather, you should try to collect eggs more often. When eggs freeze, the shells crack, and the egg is ruined.

Check Eggs for Freshness

There is a little air sac in one tip of chicken eggs that begins to seep out after the egg is laid. You have probably noticed this air sac when you peel hard-boiled eggs. The unshelled egg is not completely egg-shaped; there is a flat spot at one end. This air sac is helpful when testing eggs for freshness.

To check for freshness, fill a large, clear bowl or pitcher with water. Gently place an egg into the water, taking care not to crack the shell. A new, fresh egg will lie flat on the bottom. If the egg is a few days to a week old, one end of the egg will bob up. There might be a small tilt, or the egg may even stand on end. As long as one end stays in contact with the bottom of the bowl, the egg is still fresh enough to eat. However, if the egg floats to the surface, toss it out. It is no longer fresh enough to eat.

Educate Others About White and Brown Eggs

Depending on the breed of your chickens, the eggs you get will be white, brown, or even green or blue! Be prepared to hear all sorts of myths and falsehoods about the color of eggs and learn the facts so you can educate others about white and brown eggs.

Photo Courtesy of Kristen Harrell

One of the biggest pieces of misinformation you will hear is that brown eggs are healthier and more nutrient-packed than white eggs. People will tell you that brown eggs are better for you and better for baking than white eggs. These are not true. There is absolutely no nutritional difference between a white egg and a brown egg. One is not better for baking or boiling than the other. The inside of a brown egg and a white one is identical. The only difference is the outward color.

Some breeds of chickens lay eggs that are light blue or green. I have had friends assume the green eggs are rotten. Even after I assured them that the eggs were fresh—laid that morning!—they were still skeptical.

The eggs from your suburban flock will be fresher, tastier, and more nutritious than store-bought eggs. The yolks of store-bought eggs are pale yellow, but the eggs you will get will have bright yellow or orange yolks due to their freshness and your chickens' diet.

Commercial egg farms traditionally keep the hens confined in large barns. They are well-fed but are not free to forage for insects and fresh greens. Your hens will eat grasshoppers, grubs, and aphids, along with seeds, sprouts, and grass. All these additions to your hens' diet improve the taste and quality of the eggs you get from your suburban flock.

Gift Your Eggs

From time to time, you may find yourself swimming in eggs. In many suburban communities, local ordinances prohibit selling the eggs for commercial gain. Consider giving a few cartons of eggs to your neighbors as a gesture of goodwill. This is a great way to thank your neighbors for putting up with your hens.

The Broody Hen

As a suburban chicken farmer, you're probably aware of the joys of raising chickens for eggs, entertainment, and companionship. Hens are efficient egg producers, and, in most cases, they don't have a particularly strong maternal instinct. They are unaffected when you collect their eggs. But occasionally, a hen will go broody. Let's look at why this is and what you can do about it.

What Is a Broody Hen?

A broody chicken is a hen that has gone into a special state of mind, preparing to incubate and hatch her clutch of eggs. This behavior is usually seen in hens that are at least a year old, and it involves several behavioral changes that are quite different from a normal chicken.

When a hen goes into a broody state, she will stop laying eggs and instead focus all her energy on sitting on a nest. The broody state can last for several weeks, and the hen will be very protective of the area around her nest. She may aggressively attack people or other chickens that come too close to her nest. The hen may also become very vocal, clucking loudly and frequently to protect her nest. Additionally, the hen may become more reclusive, spending most of her time in the nest and only coming out to eat or drink.

Why Do Hens Become Broody?

A hen's broody behavior can be triggered by a number of different factors, including hormones, age, stress, the presence of eggs in the nest, or a change in the environment. Generally, older, more mature hens are more likely to become broody. Their bodies release a hormone called prolactin that triggers the urge to try to hatch eggs. It is an instinctive trait. But there is one problem; unless you have a rooster, the eggs are not fertilized. The broody hen is trying to hatch dud, unfertilized eggs.

Breaking a Broody Hen

A broody hen will sit on her unfertilized eggs so long that she puts her own health at risk. She will not eat or drink as much as she should. One hen's broodiness can spread to the other hens in your flock. This will drastically decrease the number of eggs you will get for the next several weeks. Here are some tips for breaking a broody hen.

Move the Broody Hen Off Her Nest

Remove the broody hen from her nest and take her outside. She won't be happy about this, and she might try to peck you. Wear gloves to protect your hands, and gently lift her from her eggs. Offer her some treats and fresh water to encourage her to stay outside. You will need to do this more than one time. It may require you to physically remove her from her nest several days in a row.

Block Off the Nesting Boxes

Temporarily close down the nesting boxes to prevent the broody hen from returning to her nest. Remove the eggs and the nesting material from the box. This makes the nest seem unwelcoming to the broody hen. You can also close the chicken coop door during the day so the broody hen cannot go back in. The drawback to this, of course, is that your other hens will not have access to the coop or the nesting boxes either.

Try a Frozen Water Bottle

Admittedly, this is a tip I have never tried, but I have heard that some people have had success breaking a broody hen by using a frozen water bottle. First, remove the eggs from her nest, then slip a frozen water bottle under the hen. The theory is that the cold temperatures on her underside will be a signal to her that the eggs are cold and not viable.

Put the Broody Hen in Time Out

If you still can't convince the broody hen to leave her nest, you can try to put her in a cage or a crate for a few days. Make sure she has access to fresh water and plenty of food, but keep the crate free of bedding material. After a day or two, let her out, and watch what she does. If she immediately returns to her nest, she is still broody. Put her back in timeout and test her again the next day. If she is a stubborn hen, it may take several days to break her of her broodiness.

Without the eggs beneath her and access to nesting material, the broody hen's triggers will ease, and her hormones will regulate themselves. The broody hen will return to the flock and eventually continue her egg production.

Chapter 11 Summary

In this chapter, we covered the process by which hens produce eggs that they lay approximately once a day. When the conditions are right—the hens are housed in a secure place, well-fed, adequately hydrated, and stress-free—your suburban hens will begin laying when they are between 18 to 22 weeks old and will continue to lay for several years. We explained the link between light and egg production. In addition, we presented tips for collecting and storing eggs. We learned that you shouldn't wash the eggs because it removes the natural bloom on the eggshells. Lastly, we talked about broodiness in hens and how to break a broody hen, so her behavior won't have a snowball effect on your whole flock.

Happy, well-cared-for hens will keep your family supplied with fresh, tasty eggs and allow you to offer cartons of eggs to your friends and neighbors.

CHAPTER 12

Breeding Chickens

One of the joys of having suburban chickens is raising them from tiny, fluffy chicks into mature, personable hens. As you get more involved and experienced in raising backyard chickens, you may consider breeding your own hens and hatching your own chicks. Although most hobby chicken farmers never reach this step and are happy to replenish their flock with chicks from their local feed store, breeding your chickens can be a rewarding experience. In this chapter, let's look at the process of breeding your own hens in a suburban setting.

The Reproductive Anatomy of a Chicken

Before we can get into the nitty-gritty about chicken breeding, it is important to have an understanding of the reproductive anatomy and biology of chickens. Chickens are highly social animals, and their reproductive behavior is closely linked to their social interactions within the flock. Here is a brief recap of the reproductive biology of a chicken.

A hen's reproduction system is really a two-part system consisting of the ovary and the oviduct. Female chicks are born with two ovaries, but the right one does not continue to develop after the chick hatches. Only the left one does. It becomes fully developed when the hen reaches maturity.

The hen's ovary is located along her back, about at the midway point between her head and tail. Within the ovary are thousands of minuscule ova, which will begin to develop into eggs one at a time. A sac within the ovary produces a continual amount of yolk material. The yolk material, encased in a membrane to hold it all together, passes from the hen's ovary to her oviduct.

When a rooster mates with a hen, the hen stores the male's sperm in a specialized organ within the oviduct called the sperm storage tubule. The sperm from one mating can be stored for two or three weeks and will be used to fertilize all the eggs the hen lays during that time period. It is not necessary for the rooster to individually fertilize each egg.

In the oviduct, the albumen, the clearish white part of the egg, is added to the yolk, and the entire thing is encased in the eggshell. The entire process takes roughly 24 hours, which is why a hen will lay about one egg per day.

Fertilizing the Hen

The hen must have stored sperm from a rooster to lay fertilized eggs that will become chicks. There are two ways that a hen can become fertilized, either through natural mating with a rooster or by artificial insemination.

If you have a rooster in your flock, a rarity for suburban chicken farmers, then you know that the male needs no encouragement from

you to mate with the hens. He is just doing what nature intended. As noted earlier, the mating process itself is fairly quick, but it can sometimes be brutal. The rooster stands on top of the hen's back, digs his claws into her, and aggressively rubs against her. The action can break off her feathers and leave her bloody. Once he has mated with a hen, she will remain fertile for a few weeks, so while there is no need for him to continue mating with her but he will probably continue to do so anyway.

In the absence of a rooster or in cases when you want to strictly control the genetics of chickens, it is possible to artificially inseminate hens. You first need to acquire semen from a rooster, which, as it turns out, is surprisingly easy. You can order it on Amazon or from a farm and fleet store.

The insemination process is rather simple. Hold the hen upright and keep her wings down with your arm. Apply pressure to the vent area of her abdomen. This will cause the oviduct to protrude slightly. Then you just need to introduce the semen using a bulb syringe. It is recommended that you inseminate the hen twice in the first week and once a week thereafter.

Factors to Consider Before Breeding Your Chickens

Although incubating and hatching your own chicks might sound like a great project, there are some things to consider before you start. You need to make sure you are following good breeding practices, that you have the space needed for incubating and raising chicks, that you have the financial resources to care for them, and that you know what you plan to do with the chicks you hatch.

Breeding Practices

FUN FACT

How Many Chicks?

According to the USDA, 9.39 billion eggs were produced in the United States in March 2022. In the same month, 61 million egg-type chicks and 845 million broiler-type chicks were hatched. Chicks can be hatched year-round, but the most popular time to breed chickens is from February to May.

There are many different breeds of chickens that have been carefully developed over the years by expert breeders. Through careful and selective breeding practices, they have been able to develop the most ideal traits for each breed. When you breed your own suburban chickens, be a responsible breeder that strives to maintain the integrity of the breed.

Think of breeding chickens like you would dogs. There are many different breeds of dogs. At one extreme, you have responsible dog breeders who take great pains to produce quality pups that are a credit to their breed, and then there are irresponsible "backyard breeders" at the other end of the spectrum who don't pay attention to genetics or good breeding practices while they churn out mixed-breed mutts. The mutts might be cute, but they take away from the breed qualities. When breeding your hens, take the time to find suitable breeding stock so you can maintain the integrity of the breed instead of producing mixed-breed chicks.

Space and the Number of Chicks

Many communities have rules, laws, and ordinances dictating how many chickens you can keep at one time. If you incubate and hatch too many chicks, will you be exceeding your chicken limits? You need to consider this before you jump into chicken breeding.

Photo Courtesy of Madison Woodard.

You also need to consider your own space. You will need a dedicated place to set up your incubator and a brooder box. This needs to be a secure, quiet, dry, warm spot. An unheated garage or shed is not ideal for this. Do you have a place in your home where you can set up the incubator and brooder box? And do you even want this in your house? There are odor and noise issues you need to keep in mind when deciding if you want to try your hand at breeding your hens.

Time and Resources

Incubating eggs and raising chicks is not necessarily time-consuming, but there are tasks that need to be done on a consistent basis and in a timely manner. For example, eggs in an incubator need to be turned five times a day at regular intervals. Before you decide to tackle this project, you need to make sure someone is available throughout the day to turn the eggs. It only takes a few minutes to do it, but it is not something that can wait until you get home from work.

In addition to time, you need to take a look at your financial resources to make sure you can afford to purchase the equipment you need, as well as the food for the new chicks. You can't skimp on proper feed and supplies.

What Do You Plan to Do with the Chicks?

Prior to investing in the equipment you need to breed your suburban hens, you need to think about what you want to do with the chicks you will be hatching. If you are incubating and hatching the chicks for your own use as a way to replenish your flock, you will need to keep it small. It is easy to get carried away or to want to incubate a few extra eggs just in case some don't hatch, but don't go overboard. You could end up with way too many chicks for your backyard.

Perhaps you plan to sell the chicks to other suburban chicken farmers. First, make sure that you are allowed to do so in your community. Some areas have laws prohibiting backyard chicken farmers from turning their hobby into a commercial business. This means you cannot legally sell your eggs or the baby chicks you hatch. You can, however, give your chicks away or trade them with others. Back when my kids were active in 4-H, there were a few families that hatched their own chicks, then traded chicks with each other. This gave 4-Hers a few different chicken breeds to show at the county fair.

One other thing to keep in mind is that roughly half of the eggs you incubate will hatch roosters. Most likely, you will not be allowed to keep them once they mature. You should have a plan in place for getting rid of the roosters.

Selecting an Incubator

Whether you shop at your local farm and feed store or look online, you will find that there are several incubators to choose from. With so many options on the market, it can be hard to decide which is best for the job. Here are some features to look for when making your buying decision.

Temperature Control

One of the most important considerations when buying an incubator is its temperature control. The ideal temperature for hatching eggs is 100.5°F. To get the best results, you'll need an incubator that maintains a consistent temperature and has an adjustable thermostat. This will ensure that your eggs are incubated in the ideal environment for hatching.

It's also important to look for an incubator that has a temperature alarm. This will alert you if the temperature drops or rises outside of the ideal range and give you the chance to adjust it as needed.

Humidity Control

Humidity is just as important as temperature when it comes to hatching eggs. The eggs need to be kept at a consistent level of humidity in order to hatch successfully. Look for an incubator that has a built-in hygrometer and adjustable humidity control. This will allow you to keep the humidity levels optimal for hatching, between 50 and 55% for the first 18 days and 70% for the remaining days.

Automatic Turning Trays

Turning the eggs regularly is essential for successful chick development and hatching. The eggs need to be turned between three and five times every day for the first 18 days of incubation. This can put a wrench in your schedule. To take the responsibility of remembering to turn the eggs off you, look for an incubator that comes with a turning tray. The turning tray should be able to move the eggs in a circular motion and turn them at regular intervals. You should also look for an incubator that has a timer for the turning tray. This will ensure the eggs are turned at the same intervals and will reduce the risk of over-turning them.

Proper Ventilation

Although it is important to maintain the temperature and humidity level, an incubator should also have proper ventilation. Look for an incubator that has adjustable vents and a fan to circulate the air. This will keep the eggs at the correct temperature and humidity levels. It will also help prevent the eggs from developing mold or bacteria. It's also a good idea to look for an incubator that has a built-in air filter. This will help keep the air inside the incubator clean and free of contaminants.

Incubating Eggs

Incubation is the process of providing conditions for eggs to develop until they hatch. This includes maintaining the temperature and humidity of the incubator, as well as turning the eggs regularly. Since baby birds do not completely develop within their mothers' wombs, like mammal babies, incubation is necessary to allow the chicks to grow and develop to the point when they are ready to enter the world. Incubation can be done naturally, with a mother hen sitting on the eggs or artificially, using an incubator.

Natural Incubation

A hen needs to be broody to incubate her eggs. Because broodiness is not a quality people look for in backyard hens, chicken breeders have tried to breed the broodiness out of their chickens. You may discover that your hens lack the strong maternal instinct to incubate their own

Photo Courtesy of
Kelly Kneeland

eggs. You can help trigger the hormone that causes broodiness in hens by providing clean, quiet nesting boxes with plenty of nesting material. Don't collect the eggs. Seeing a clutch of eggs—and feeling them under her—will often trigger the release of hormones needed for the hen to instinctively sit on her eggs. Once she starts incubating her eggs, the hen will not want to leave the nest for very long. Set up a supply of food and water nearby so she can grab a quick bite to eat and go back to her incubation duties.

When you allow your hens to incubate their eggs naturally, there isn't much more for you to do. The hens can handle it. Mark the calendar, though. Chicken eggs hatch 21 days after incubation starts.

Using an Incubator

First, thoroughly wash and sanitize your incubator and set it up in a warm, quiet spot where it won't get bumped or knocked over. Your incubator will come with instructions on how to set the correct temperature and control the humidity in the incubation box. The ideal temperature is 100.5 °F, and the humidity should be between 50 and 55%. Get the incubator set up and at the correct temperature before you set the eggs.

Some incubators have automatic turners, but if yours does not, you need to set a turning schedule for the eggs. They need to be turned or rotated between three and five times every day for optimal chick development. The yolk, albumen, and embryonic chick within the eggshell are subject to the law of gravity. If the egg is not turned, the embryo will settle against the shell, and the development of the chick will be negatively impacted. When a hen incubates her eggs naturally, she will turn them herself. She knows instinctively to do this.

Before you manually turn the eggs, make sure to wash your hands or put on gloves. Germs, dirt, oils, and lotions on your hands can pass through the eggshells to the developing chick. Work quickly but carefully to turn all the eggs and securely close the incubator lid again. This will keep as much of the heat and moisture in as possible.

On day 18 of the incubation process, the chick is fully developed and preparing to hatch from its shell. You can stop turning the eggs at

this point. Arrange all the eggs so that the large end is pointing up. Keep the temperature in the incubator at 100.5°F but increase the humidity level to 70%.

On the 21st day, the eggs should start to hatch. Watch and enjoy this process all you want, but resist the urge to help. There may be tiny blood vessels still connecting the chick to the shell that need to dry up before the chick can fully emerge. If you try to hurry the process by helping a chick break free of its shell, you might inadvertently cause the chick to bleed. Be patient. While most chicks make their escapes in five to seven hours, others can take up to 24 hours to completely hatch.

Keep the newly hatched chicks in the incubator until the other eggs have hatched. Often, the sound of a chick peeping is encouragement for other chicks to break out of their shells.

If the 21st day has arrived and nothing seems to be happening with your eggs, give them another day or two. It could be that the incubation process didn't start exactly when you thought it did. Or, when you opened the incubator to turn the eggs and the temperature dipped, it may have slowed the process. By day 23, if there are still no signs of hatching, you can try candling an egg—holding it up to a bright light to see silhouettes inside the shell. You will be able to tell if there is a chick still forming inside or if the egg was a dud that never developed a chick.

Eggs in Suspended Animation

While it is true that it takes chicken eggs 21 days to incubate and hatch, exactly when the incubation begins is an inexact science. Hens typically lay one egg a day, but they don't start to incubate that first egg on the first day. Instead, they wait until they have a small collection of eggs (a clutch) before they begin the incubation process. That means that some of the eggs are several days old before they are incubated. What is going on during those days? Is it possible to incubate an egg that is already a few days old?

After they are laid, the eggs go into a temporary state of suspended animation. Think of it as being in a holding pattern while they wait for their siblings to be laid. In the wild or in your backyard chicken coop, the mother hen will not sit on these eggs to warm them, but she will turn and rotate them. When she has a good number of eggs ready, she will begin the incubation process. The countdown begins for all the eggs, regardless of when they were laid.

When you collect the fertilized eggs from your hens, you don't have to start incubating them immediately. Store the eggs in a cool, dry place with the small end of the egg pointing down. Twice a day, turn and rotate

the fertilized eggs. Then, when you have enough fertilized eggs, you can put them in the incubator and hatch them as one batch.

After the Chicks Hatch

After the chicks have hatched, they can be moved to their brooder box. Follow the same guidelines that were discussed in Chapter 3. That information applies equally to chicks you acquire from an outside source and ones you hatch yourself. Feed the newly hatched chicks a good-quality chick starter feed, offer them plenty of fresh water, and keep them warm with a heat lamp. They will continue to grow and thrive, and soon, they will be ready to take their place in your backyard coop.

Chapter 12 Summary

Most suburban chicken farmers do not breed their own chickens and incubate their own eggs, but doing so can be a fun, rewarding, and educational activity. Before you make your decision about breeding your own hens, you need to consider how much time, financial resources, and physical space in your home you have available for this project. You will also need to know what to look for when buying an incubator and how the incubation and hatching process works. Lastly, you should have a plan for what you intend to do with the chicks you hatch, knowing that about half of them will be roosters. For a suburban chicken farmer, small-scale incubation can provide a way to replenish your flock and give you and your family an opportunity to marvel at the wonders of life from a chicken's point of view.

CHAPTER 13

Health and Wellness of Your Flock

The health and well-being of your suburban chickens should be a priority. You will need to take steps to reduce the risk of your hens getting injured or being exposed to contagious diseases. Identifying signs of sickness in chickens, especially in the early stages, is crucial for maintaining the health of your entire flock and preventing the spread of disease. In this chapter, we will discuss poultry health, how to monitor the health of your flock, common chicken diseases, and first aid for chickens. We will also offer information about avian influenza, also known as chicken flu, so you understand the risks to your birds and your family and can differentiate facts from fiction.

Recognizing Signs of Illness in Chickens

> Chickens must bathe in dusty sand regularly to keep themselves free of mites. This dust-bath area needs to be separated from the regular chicken run, which will fill with bedding and poop quickly.
>
> ELIZABETH SORBY

Your hens can't tell you they are feeling unwell. They can't articulate that they have a headache or feel nauseous or are fatigued. But they will exhibit some signs to let you know that something is not right. Your job is to recognize those signs and understand what they might mean. What are some of the signs of illness in your suburban chickens? Here are a few things to look for.

1. Behavior Changes

One of the first signs of illness in chickens is a change in behavior. Chickens that are sick may become lethargic, less active, and less interested in food and water. They may also separate themselves from the rest of the flock, sit in one place for long periods, and appear to be uncoordinated. Keep an eye on your chickens and note any changes in their normal behavior.

Chickens that are sick may also exhibit aggressive behavior toward other chickens or humans. This is a defense mechanism and a sign that they are not feeling well.

2. Respiratory Symptoms

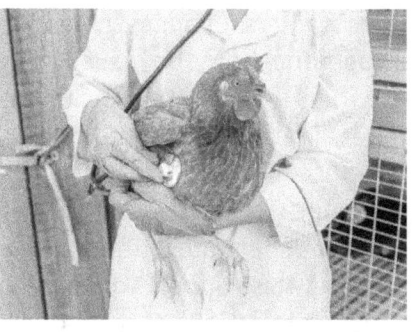

Respiratory symptoms, including coughing, sneezing, wheezing, and labored breathing, are common signs of illness in chickens. Birds have a rapid breathing rate, making them more susceptible to airborne illness. Respiratory symptoms can be caused by a variety of illnesses, including infectious bronchitis, avian influenza, and mycoplasma. If you notice any of these symptoms in your chickens, it is important to isolate them from the rest of the flock immediately to prevent the spread of disease.

3. Changes in Chicken Droppings

Changes in feces can also be a sign of illness in chickens. Healthy chicken droppings should be firm and brown with a white cap. Green or yellow droppings can be a sign of diarrhea, which can be caused by a variety of illnesses, including coccidiosis and bacterial infections. Blood in the droppings can be a sign of internal bleeding or injury.

4. Abnormal Egg Production

A sudden decrease in egg production or abnormal eggs may be a sign that your chickens are not feeling well. Abnormal eggs can include soft-shelled eggs, misshapen eggs, or eggs with irregularities in the shell. These can indicate nutritional deficiencies, stress, or illness.

5. Ruffled or Unkempt Feathers

Chickens that are sick may have ruffled, unkempt feathers or bald spots. They may also have a dull or discolored appearance to their feathers. Changes in feather appearance can be caused by a variety of illnesses, including mites, lice, and fungal infections.

6. Appetite Changes

If your hens are ill, they may lose their appetite or eat less than usual. They may also appear to be drinking more water than usual. Be sure to continue to offer them high-quality feed and fresh water, monitor their weight, and make sure they are getting enough nutrition.

7. Coordination or Movement Difficulties

Coordination issues or movement difficulties could indicate injury or illness in chickens. Look for obvious injuries, wounds, or open cuts that could be causing the bird to have difficulty walking, standing, or moving around. A hen may appear to be uncoordinated or stumble as a result of being sick.

8. Unusual Appearance of Combs and Wattles

The comb and wattles, located on top of the chicken's head and under the chin, should be bright red and free of discoloration or swelling. Texture and color changes in your chickens' combs and wattles may be a sign of fungal infections, mites, or dehydration.

Common Chicken Illnesses

Fortunately, chickens are hardy birds that enjoy good health when they are properly cared for, well-fed, and have access to fresh, clean water. Even though they are generally healthy, that doesn't mean that they are immune to illness. Here is a brief overview of some of the most common illnesses that could be an issue for your suburban chickens.

Coccidiosis

Coccidiosis is an internal parasitic disease that can affect chickens, causing diarrhea, weight loss, and a decrease in egg production. Chickens contract coccidiosis by consuming contaminated food or water, which is why it is important to keep your chickens' food and water dishes clean, provide fresh water, and ensure that your coop is clean and dry.

The good news is that coccidiosis is treatable when it is caught early. Consult with your veterinarian about the best course of treatment; however, an outbreak can usually be treated by using Amprolium.

HEALTH ALERT!
Salmonella

Chickens can carry salmonella bacteria, which can be transmitted to humans through touch or consumption of contaminated eggs or meat. Symptoms of salmonella can vary and tend to be more serious in young chickens. To minimize your risk of contracting salmonella from your chickens, it's crucial to practice good hygiene habits, such as washing your hands thoroughly after handling your birds or their waste.

Amprolium is not an antibiotic. It works by blocking the parasite's ability to multiply and spread. Amprolium is available at your local farm and feed store or can be ordered online. Amazon carries it. Powder Amprolium is easily added to your chickens' drinking water, but if you have a hen that is so sick it is not drinking or eating enough, Amprolium can be administered orally. The directions and dosages are clearly printed on the packaging.

The package label directs you to treat your suburban hens with Amprolium for seven days after an outbreak of coccidiosis, but you will likely see a drastic improvement in your chickens in a day or two. Backyard chicken farmers living in warm and humid regions may battle with coccidiosis outbreaks more frequently than others. It is possible to use Amprolium as a preventative measure to keep the coccidiosis parasite at bay.

Marek's Disease

Marek's disease, also called fowl paralysis, is a highly contagious viral disease that affects chickens and can cause a range of symptoms, including paralysis, weight loss, and internal tumor growth. It affects commercial poultry farms more often than small-scale flocks, but suburban chickens are also susceptible to the illness. Marek's disease is spread when the birds inhale the virus, which lives in feather dander. The virus itself is easy to kill. However, it can live for a long time in the feather dander, making Marek's disease somewhat difficult to control. Because Marek's disease spreads so rapidly and can be deadly, it has the potential to wipe out an entire flock.

Unfortunately, infected chickens do not recover from Marek's disease. The disease is incurable and fatal. In commercial poultry operations, the flock would be destroyed, and the facility thoroughly cleaned and sanitized to eradicate feather dander. As a preventative, large-scale commercial chicken farms vaccinate their birds. There are two points to make about this. First, it has been proven that vaccines alone are not enough to prevent an outbreak of Marek's disease. Second, vaccines for Marek's disease are not available in small numbers for suburban chicken flocks.

For suburban chicken farmers, the best advice is to take steps to prevent Marek's disease. Those include purchasing your chicks from a top-quality hatchery that follows good biosecurity practices and quarantining new chickens before introducing them into your flock. You should also avoid overcrowding in your coop as the disease spreads through close contact.

Botulism

A bacterial disease, botulism is caused by consuming contaminated food or water that contains *Clostridium botulinum*. The bacteria is found in decaying vegetation and rotting animal carcasses. For suburban chicken farmers who practice good coop hygiene and who make sure their backyard is free of hosts for the botulism bacteria, the risk of an outbreak among their hens is low.

Symptoms of botulism initially present as fatigue, reluctance to move around, and excessive sleepiness. The hallmark symptom of botulism in chickens, however, is paralysis of the neck. The bird's head will droop and bob, but as the illness progresses, the chicken will be unable to even lift its head. Sadly, the paralysis of the bird's neck muscles will progress to the point when the chicken cannot breathe. There is no treatment for botulism in chickens. The best thing you can do is to take steps to prevent an outbreak.

To prevent botulism, ensure that your chickens have access to clean, fresh water at all times and regularly clean and disinfect your chicken feeders and waterers. Since the botulism bacteria makes its home in rotting plant material and dead animals, avoid feeding your chickens spoiled or moldy food. Don't allow your chickens to drink from stagnant water sources like puddles or water that accumulate at the base of your gutter downspouts.

Immediately remove any dead animal you find in your backyard or near your coop. If one of your hens has died, dispose of the body immediately. Also, inspect your henhouse, chicken run, and backyard for other animal carcasses, including dead mice, squirrels, snakes, or wild birds. Sometimes it is difficult to spot these, but let your nose be your guide. Occasionally, your birds may be presented with an opportunity to come in contact with a larger carcass, such as a deer or raccoon that was hit by a car and wandered into your yard to die.

Respiratory Illnesses

Falling under a broad umbrella that includes several different conditions, respiratory illnesses are a common yet mostly curable problem among backyard chickens. Poor ventilation, dusty bedding, and exposure to drafts can cause a variety of symptoms, such as coughing, sneezing, nasal discharge, and difficulty breathing. The most common respiratory illness in chickens is CRD, or chronic respiratory disease, which is caused by a bacterium, *Mycoplasma gallisepticum*. Others include pneumonia, bronchitis, laryngitis, and tracheitis. All these illnesses are treated the same way.

Contact your veterinarian if you notice signs of respiratory illness in one or more of your hens. Most likely, the veterinarian will give you an antibiotic to treat the disease. Because some antibiotics can pass through

to the hen's eggs, you can ask your veterinarian to prescribe one of the antibiotics that is best for egg layers, like chlortetracycline or a combination of lincomycin with spectinomycin. Other antibiotics that are effective against mycoplasma include doxycycline, enrofloxacin, and amoxicillin;

however, you should discard the eggs you get from treated hens for a few weeks. Your vet will give you guidance on when the eggs should be good to eat again.

Respiratory illnesses can be treated with antibiotics, but prevention is key. Ensure that your chicken coop is well-ventilated, clean, and free of dust and mold. Additionally, you should quarantine new birds before introducing them to your flock to prevent the spread of respiratory illnesses. You should also avoid overcrowding your coop, as this can increase stress levels and make your chickens more susceptible to illness.

External Parasites

External parasites, such as lice, mites, and worms, can cause a range of health problems in suburban chickens, including feather loss, weight loss, and diarrhea. Chickens can also be bothered by fleas, bedbugs, flies, and ticks.

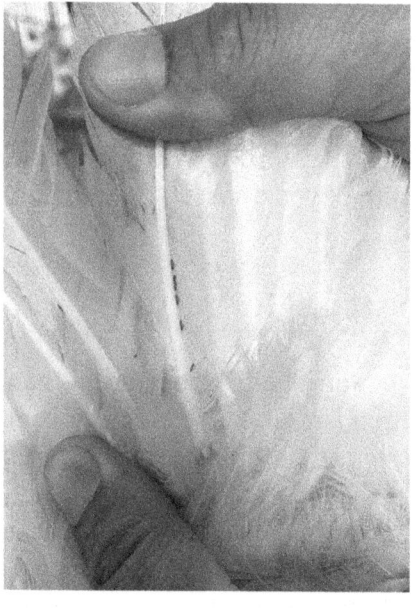

If your chickens are dealing with external parasites, you will notice some signs. For example, the hens may seem to be preening themselves excessively. They might scratch themselves, pull out their feathers, or have broken feathers. Inspect your hens regularly for parasites. Pick up a bird and flip it onto its back. Gently extend a wing and look closely at the skin, especially along the crease. Spread the feathers so you can see down to the skin. Also, inspect the vent area, the leg shafts, and the back.

Because mites feed on the blood of hens, they can cause your birds to become anemic. When this happens, you may notice that the normally bright red combs and wattles on the chicken have grown pale.

One time, we took several of our chickens to a poultry show, and they all came back with lice. We don't know if they got lice from other birds at the event or if the cages provided by the event organizer were infested, but we knew how to get rid of the lice. We used Sevin dust carbaryl, in a powder form. One by one, we dusted the chickens, thoroughly working the power through their feathers and down to their skin. We also removed all the bedding material from the coop and nesting boxes and scrubbed all parts of the coop. When it was completely dry, we poured generous amounts of Sevin dust into the nesting boxes, along the floor, and in all the nooks and crannies.

Sevin dust is sold over the counter at your local feed store. While Sevin dust did the trick for us, there are other over-the-counter products on the market that are also effective at treating mites, lice, and fleas. If you are unsure which product is best for your situation, contact your veterinarian for advice. Once again, good hygiene and preventative measures are helpful in reducing your chances of having an external parasite problem in your suburban chickens. Parasites can be prevented through regular cleaning and disinfecting of your coop, as well as providing your chickens with periodic dust baths to keep their feathers clean and healthy.

Fungal Infections

Fungal infections in suburban chickens can cause a range of symptoms, including respiratory problems, diarrhea, and feather loss. A hen can become uncoordinated or may breathe with a gurgling noise because of a fungal infection. Additionally, her droppings may be green and watery and cling to the feathers of her vent.

Three of the most common fungal infections in chickens are aspergillosis, moniliasis, and mycotoxicosis.

Aspergillosis can be a chronic condition in adult chickens. It is caused by fungus produced when wood, bedding material, and feed decay and break down. The fungal spores are inhaled by the hens, causing the infections.

Moniliasis is also called thrush. It primarily targets the upper digestive tract of chickens and is caused by a yeast-like fungus. You will notice a white thickening of the crop area, a common characteristic of moniliasis. The fungus can result in a breakdown of the chicken's gizzard. Moniliasis infects chickens through contaminated food and water, not from bird to bird.

Mycotoxicosis comes from mold growing on chicken feed that can become toxic. When suburban hens are infected with mycotoxicosis, they lose their appetites, lay fewer eggs, and the shell quality of the eggs will be quite poor. The fungal infection impacts the bird's liver, gastrointestinal tract, and immune system.

When a fungal infection is suspected, the first thing you should do is remove the source of the fungus. That is probably the chickens' feed, but it could also be the water dish or an environmental source. Many suburban chicken farmers swear by treating chickens with fungal infections with probiotics and soluble vitamins. Boosting the protein in your flock's diet will also help your hens recover quicker from a fungal infection.

To prevent fungal infections, ensure that your coop is kept dry and well-ventilated. Regular cleaning and disinfecting can also help to prevent the growth of fungi. Be sure that the chicken food you feed your hens has not spoiled and that it has been stored in a dry, moisture-proof container. If you suspect that your chickens have a fungal infection, consult with a veterinarian to determine the best course of treatment.

What You Need to Know About Avian Influenza

Avian influenza, also known as bird flu, has gotten a lot of press in the last ten years, and in many cases, it has scared people away from getting backyard chickens. They fear that raising chickens in their suburban

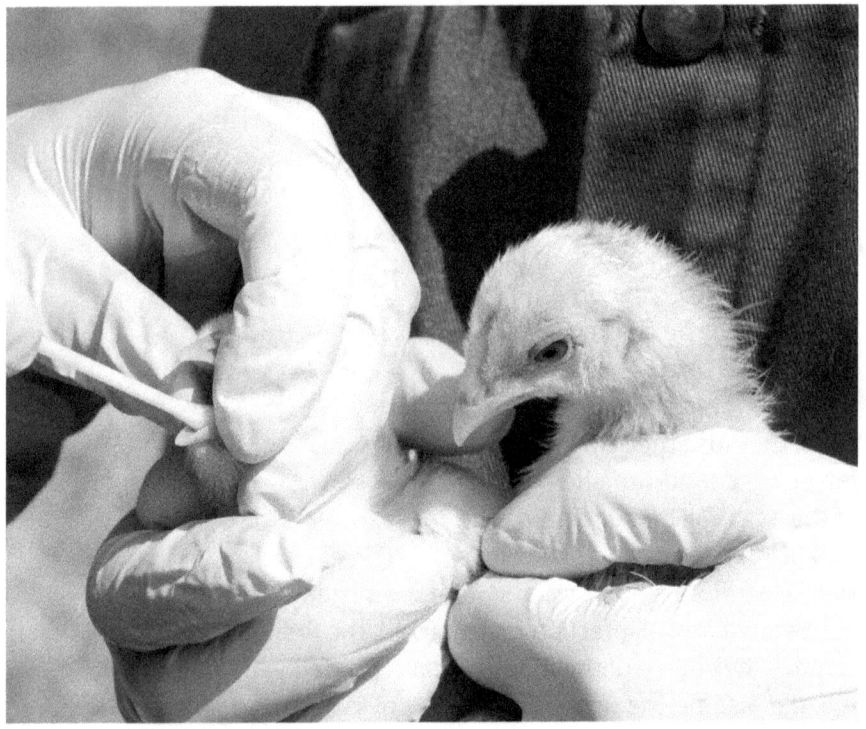

home will put them and their families at risk of catching a deadly disease. There has been a lot of misinformation and misunderstanding surrounding bird flu, and as a suburban chicken farmer, it is important to untangle the knot of information so you can understand the realities of avian influenza. Let's look at the basic facts about this illness. Armed with accurate knowledge, you will be able to calmly answer questions and concerns from your neighbors or friends.

What Is Avian Influenza?

Avian influenza is a highly contagious viral respiratory disease that can affect both wild and domestic birds, including suburban chickens. It is a type of zoonotic virus, meaning that it is carried by and mostly affects animals, but it can pass to humans. Bird flu can take two forms— low pathogenic avian influenza (LPAI) or high pathogenic avian influenza (HPAI), and numerous variants of each have been documented.

HPAI garners all the media attention because it is extremely contagious and spreads quickly through flocks. The death rate among chickens that contract this form of bird flu is more than 90%. However, it is important to note that HPAI mostly strikes large-scale, commercial poultry farms. Less than 10% of outbreaks have been attributed to small-scale, backyard hobby farmers.

HPAI is not common in the United States, despite what you may have heard. In fact, this highly deadly variety has only been detected in the United States on a few occasions, in 1924, 1983, 2004, 2015, and 2022. Each time, the outbreak was caught early and contained quickly. There have been no human deaths related to HPAI in the US, though there have been some in Asia.

By far, the most common type of bird flu is LPAI. This version of the virus is mostly harmless, with a low death rate among chickens. Infected chickens exhibit cold-like symptoms.

How Is Avian Influenza Spread?

Avian influenza is spread from bird to bird. The virus lives in the respiratory and digestive systems of infected birds and can be found in respiratory droplets, nasal secretions, saliva, and droppings. Healthy birds can become infected by breathing in the water droplets or coming into contact with infected droppings.

In a large commercial poultry farm, it is easy to see how the bird flu virus can spread among chickens housed in close quarters. For smaller, isolated suburban chicken flocks, the risk is lower because the hens are contained. They don't generally encounter other chickens. That doesn't mean they are safe from infection. Wild birds can spread bird flu to domesticated chickens. Studies have shown that water birds like ducks, geese, and seagulls are the main culprits.

What Are the Symptoms of Avian Flu?

With LPAI, the symptoms are typically mild and mimic those of a cold. The chickens will cough, wheeze, and have nasal discharge. The symptoms for chickens infected with HPAI, however, are much more severe. The birds will be in obvious respiratory distress, gasping, coughing, sneezing, panting, and making gurgling noises when they breathe. You will notice drainage from their eyes and noses, a blue pallor to their necks and throats, diarrhea, and loss of appetite. HPAI progresses quickly. Chickens that seem fine in the morning can be dead by evening.

Treating Avian Influenza

There is no cure for avian influenza, even the mild LPAI form. Sadly, LPAI can sometimes evolve into HPAI. Because of this potential, entire flocks are culled to stop the spread of the virus. Other flocks in the area, including suburban hens, will be quarantined and observed for signs of illness to make sure that wild birds haven't spread the virus.

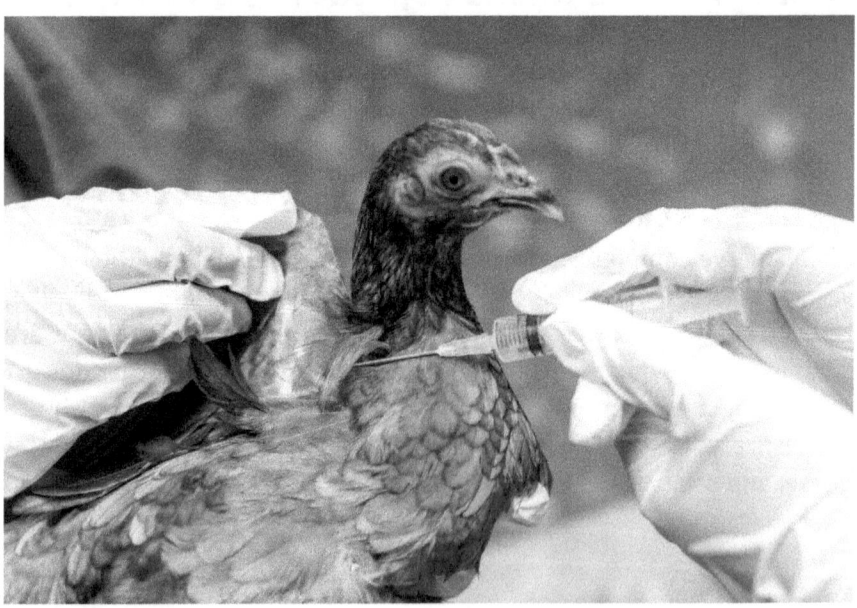

How Do I Protect My Flock from Avian Influenza?

Keep your suburban chickens isolated from other chickens as much as possible. As fun as it might sound to you, forego chicken playdates with your friends or neighbors. If you plan to enter your chickens in a county fair or area poultry show, find out what precautions the show coordinators are taking to keep the birds healthy. Avoid participating in a poultry show if there have been reports of avian flu outbreaks in the region.

The biggest threat to your suburban chickens in regard to avian influenza is wild birds. It is nearly impossible to keep wild birds out of your yard, but there are some steps you can take to reduce the risk of your hens coming into contact with geese or ducks or other wild birds. You can keep your chickens in a chicken run that is enclosed on the top. If you have a backyard pond that attracts geese and ducks, keep your hens away from it and diligently clean up the goose droppings.

In addition, you can practice strict biosecurity measures, including quarantining new birds for ten days before introducing them to your flock. Keep your coop and chicken enclosure clean, and remove droppings often. Wash and sanitize feed and water dishes on a regular basis. Practice good hygiene by thoroughly washing your hands before handling the chickens, the eggs, and the feed, then wash your hands again when you are done. Keep a bottle of hand sanitizer near your coop and use it after you rake the droppings and after you lock the hens in at night.

My husband and I each have a pair of rubber rain boots that we only wear when doing chicken chores. We don't wear any other shoes to the chicken

Photo Courtesy of Karisa Nigon

coop because we could step on chicken poop that we then transmit to other places. The rain boots stay in the garage, never in the house. Lastly, try to keep visitors to your chicken coop to a minimum.

What Are the Risks of My Family Catching Bird Flu from My Chickens?

The virus that causes avian influenza primarily targets birds, but it can morph into a form that can infect humans. As I noted earlier, there have been no cases of humans in the United States falling ill from avian influenza, although cases have been documented in other parts of the world.

If you are concerned about possibly getting sick, you can wear a mask when you do your chicken chores and when you clean out the coop. This will drastically reduce the contaminated particles that you inhale. Wear gloves when you will be coming in contact with chicken feces. Wear old clothes when you clean the coop, and immediately put them in the washing machine when you are done.

What Is the Biggest Misconception about Avian Influenza?

One common and persistent belief is that humans can catch avian influenza from eating eggs. This is not true. You cannot catch bird flu from eating properly cooked eggs, nor can you get it from consuming cooked chicken from disease-free birds. Food industry safety measures are in place to prevent contaminated meat from reaching the food supply, so you don't have to worry about catching bird flu from a package of drumsticks you buy at Costco.

Testing and Reporting Avian Influenza

Contact your local county extension office, your veterinarian, or the USDA to obtain a free avian influenza test kit. If your suburban chickens have contracted avian flu, you are required by law to report it to the authorities.

Finding a Chicken Veterinarian

The best time to find a veterinarian in your area who treats chickens is long before you need one. When you begin to build a chicken coop, purchase supplies, and set up a brooder box for new chicks, you should also start looking around for veterinarians in your area who will treat chickens. Not every veterinarian does. Most small animal vets stick to dogs and cats, with occasional rabbits and guinea pigs thrown in. Poultry—well, that's a whole new animal.

I started my search for a chicken doctor at my dog's veterinary clinic. I was pleasantly surprised to learn that there was one veterinarian on staff who was experienced at treating chickens. I got lucky.

When you start your search, I suggest you contact your current veterinarian if you have one. Since your dog or cat is already a patient there, it will be easy to add a few more patients to your account. If your dog's vet doesn't treat chickens, perhaps he or she could recommend one that does. You can also do a quick Google search. I was able to quickly locate a county-by-county list of poultry veterinarians in my state. I found similar lists for other states too. Don't trust everything you find on the internet; contact the veterinarians on the list for your county to confirm that they do, in fact, treat chickens. If they do, find out what you need to do to get established as a patient there—your hens, not you.

First Aid for Injured Chickens

As we discussed in previous chapters, chickens are prey animals; therefore, they tend to hide their injuries and illnesses to avoid being targeted by predators. Familiarize yourself with your chickens' physical characteristics, such as the color of their comb and wattles, their weight, and their usual behavior, to detect any changes that may indicate an injury or illness.

Chickens have a high pain tolerance, and they may continue to eat and drink even when injured. Inspect your chickens' bodies regularly, especially their legs, wings, beak, and vent, for any signs of injury or abnormality.

It's crucial to establish a first aid kit for your chickens, including items such as gauze pads, antiseptic spray, scissors, tweezers, and saline solution. You should also have a separate, clean, and dry area to isolate injured chickens and monitor their recovery.

Wounds are a common injury in chickens, especially in their legs, wings, and vent area. If you notice an injury, remove the bird, and clean the wound with saline solution or warm soapy water. Apply antiseptic spray or ointment to the wound and cover it with a gauze pad or bandage. Be sure to change the dressing regularly and monitor the healing process.

If the wound is severe, such as a deep cut or puncture, seek veterinary care immediately. The vet may need to clean and stitch the wound, administer antibiotics, or give pain medication to the chicken.

Broken bones are a severe injury in chickens, and they may require veterinary care. If you suspect that your chicken has a broken bone, isolate the injured chicken, and avoid handling it as much as possible. Provide a comfortable and warm environment for the chicken to rest and recover.

If you're familiar with chicken anatomy or are experienced in

Photo Courtesy of John and Kim Anderson

basic first aid, you may be able to splint the broken bone using a thin wire or popsicle stick. However, if you're not confident in your ability to splint the bone, you should seek veterinary care immediately. The vet may need to X-ray the chicken to diagnose the extent of the injury and administer pain medication or antibiotics.

Heat stress is a common issue in chickens, especially in hot and humid environments. If you notice that your chicken is panting, lethargic, or has a droopy comb, it may be suffering from heat stress. Isolate the chicken and provide a cool and shaded environment with plenty of fresh, clean water. Consider putting a frozen water bottle in its box to help reduce the temperature.

Avoid handling or stressing the chicken as much as possible and monitor its behavior and body temperature. If the symptoms persist or worsen, seek veterinary care immediately. The vet may need to administer fluids or other medication to treat the heat stress.

Occasionally a hen may experience egg binding, a relatively common issue in chickens, especially in older or overweight hens. If you notice that your chicken is straining or has a swollen cloaca, it may be suffering from egg binding. Isolate the chicken and provide a warm and quiet environment.

Photo Courtesy of
Anne Greenwood

Offer the chicken a calcium supplement or eggshell to help it pass the egg. If the chicken is still unable to lay the egg, seek veterinary care immediately. The vet may need to manually remove the egg or administer medication to help the chicken pass the egg.

Preventing Injuries in your Suburban Chicken Flock

The best way to deal with injured chickens is to prevent injuries and illnesses in the first place. You can prevent injuries by providing a safe and clean environment for your chickens, such as a secure coop and run, clean bedding, and fresh, clean water and food. You should also inspect your chicken's environment regularly for any hazards or potential sources of injury. Snip off protruding wires and pound flat any sharp nails.

Keep dogs and cats away from your backyard chickens, as they can also cause injuries.

Chapter 13 Summary

As we have discussed throughout this chapter, the best thing you can do to maintain the overall health and well-being of your chickens is to follow best practices regarding nutrition and hygiene—offer your suburban chickens a quality diet, provide clean, fresh water, and routinely disinfect the coop. When you raise suburban chickens, you are taking on the responsibility of keeping them in good health. If you take good care of your suburban chickens, you will be taking a huge step to ensure their continued good health and well-being.

CHAPTER 14

Sustainable Practices for Suburban Chicken Farming

It's important to understand the environmental impact of suburban chicken farming. Chickens produce waste, and if not managed properly, this waste can pollute the environment. Additionally, even small-scale chicken farming can contribute to deforestation, soil degradation, and greenhouse gas emissions. By adopting sustainable practices, you can minimize these impacts and create a more sustainable backyard chicken farm.

Reduce Food Waste

Food waste can be a major issue for suburban chicken farmers. Chickens will eat many things, but it's important for you to feed them a balanced and nutritious diet. By feeding your chickens the right amount of food, you can reduce waste, prevent spoilage, and save money on feed costs. Some suburban chicken farmers supplement commercial layer pellets or mash with kitchen scraps. However,

HELPFUL TIP

Chickens and Compost

Chickens can be excellent contributors to garden composting. Their bedding materials, such as straw or untreated wood shavings, combined with their manure, create a nutrient-rich material with high nitrogen content. The natural propensity for chickens to scratch and peck at the ground can also be directed toward your composting efforts. Place your compost heap inside your chicken run and let your chickens turn the pile.

it's important to make sure these scraps are safe for chickens to eat, are not spoiled or decaying, and do not contain any harmful chemicals or toxins. If your hens do not eat all the kitchen scraps, remove them the next day and discard them. It is unhealthy for your chickens to consume decaying food, and, as we saw in the previous chapter, rotting material in close proximity to your chickens increases the risk of illness.

Reduce Water Waste

Don't skimp on water for your suburban flock. Your hens should have round-the-clock access to fresh, clean water without restrictions. It is possible, however, to provide your chickens with so much water that you end up

dumping out most of it every day when you refill their water container. To conserve water, try to find the right amount, plus a bit extra, to give them.

Many chicken waterers hold three or five gallons of water, but do your hens really need that much water? In general, an adult chicken will drink about a pint of water per day. That means if you have six hens in your flock, they will drink about three-fourths of a gallon each day. You never want to allow the waterer to run dry, but you can purchase a gallon-size chicken waterer for your suburban hens rather than a three- or five-gallon one.

When you refill the chicken waterer, water your flower beds or container garden with any unused water instead of dumping it on the ground.

Use Repurposed Building Materials

When building your chicken coop and run, try to use repurposed materials. This can include reclaimed wood or recycled plastic. An unused shed, a playhouse, scrap wood, and items you pulled out of your house

during remodeling projects can all find new life as chicken coops. The door of our chicken coop used to be the front door of our house, and the window in the coop was one from our house that we saved when we replaced it with a larger window.

Check out Pinterest for ideas and tips on creating chicken coops from reclaimed building materials and other items. I am amazed at what people can create from discarded wooden pallets. Using sustainable materials will drastically lower the cost of constructing a chicken coop for your suburban chickens, especially with the increased costs of lumber and building materials. Reclaimed and repurposed building materials will also reduce your environmental impact and create a more eco-friendly chicken farm.

Use Compostable Bedding Litter

Consider using bedding material in your chicken coop that is highly compostable, such as straw, wood shavings, or hemp litter. Chicken poop is high in nitrates and makes an excellent fertilizer for your flower beds and vegetable gardens. If you have space for a compost pile and one is allowed in your neighborhood, spoiled bedding material from the chicken coop can be added to the compost bin. It will decompose quickly, and the chicken droppings will enhance the soil. The biggest plus, however, is that the dirty bedding material won't end up in a landfill.

Use Natural Cleaning Products

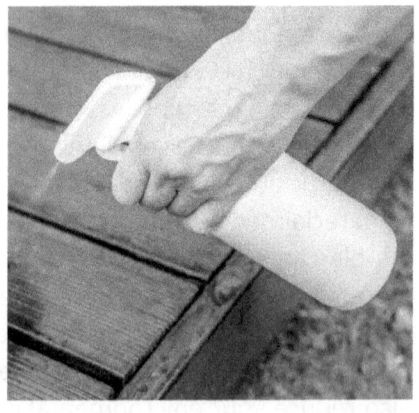

Safer for you, safer for the chickens, and safer for the environment, natural cleaning products will get the job done without harsh chemicals. You can even make your own cleaning supplies by mixing one part vinegar with water. This mixture can be used to mop the coop floor and to clean the roosts and nesting boxes. It is safe, effective, and eco-friendly.

Harness the Power of the Sun

Photo Courtesy of John and Kim Anderson

You can reduce energy costs associated with your chicken coop by harnessing the power of the sun. One way to reduce energy use is to use solar-powered lights in and around your coop and run. This will not only reduce your energy consumption but also save you money on electricity bills.

When you build your hen-house, include windows and skylights positioned to catch the sun's rays during the winter months. On sunny winter days, the coop will become like a greenhouse as the heat of the sun warms the inside.

Control Pests with Natural Methods

Chicken feed, unfortunately, attracts pests like mice, rats, and ground squirrels. Don't control the pests with chemical mouse poisons. This can be a problem for your hens, as well as for the environment. Instead, use natural or non-chemical ways to keep the rodent population down. You can use mouse traps as long as they are placed away from the chickens, so they are not injured by one. Personally, we use the oldest form of mice control—a cat. During the day, Mabel the cat prowls the yard, viciously hunting for mice. At night, she turns into a cuddly lap cat. She gets the job done!

It is also possible to control fleas, mites, and lice using all-natural, non-chemical products that are environmentally friendly and safe for your chickens. One of the most commonly used products is diatomaceous earth, a white powder made from crushed and pulverized sedimentary rocks. You can dust your chickens with the powder and sprinkle it around their chicken coop without worrying about harming your hens or the environment.

Some herbs and essential oils are also effective at repelling insects and pests. These herbs not only smell great but also have natural pest-repelling properties.

Photo Courtesy of Lana Plashchynskaya

Put Your Chickens to Work

Chickens can be beneficial in your yard and gardens. They love to forage for tasty insects, worms, and grubs—pests that can cause damage to your flowers or vegetable plants. Just keep an eye on your hens to make sure they don't eat your plants or peck at the tomatoes. As the hens are wandering around your yard, they will scratch up the grass and aerate the soil. They will also leave their droppings around, which make excellent fertilizer.

Chapter 14 Summary

Keeping a small flock of hens is one of the hallmarks of sustainable homesteading. It is possible, with some conscientious planning, to incorporate environmentally friendly practices into your suburban chicken adventures so you minimize energy consumption and reduce your carbon footprint. Using repurposed material for a henhouse, being mindful of the amount of food and water you provide, selecting compostable litter material, and making your own non-chemical cleaning supplies are all simple steps you can take to incorporate sustainable practices into your suburban chicken farm.

CHAPTER 15

Resources and Support

A s you begin your suburban chicken adventures—and even if you are a veteran chicken farmer—keep in mind that you do not have to go it alone. Just as hens are social creatures that thrive in a flock setting, you can surround yourself with a flock of other suburban chicken enthusiasts and develop a thriving community of like-minded people who will be a valuable source of support and information. In addition, there are organizations, agencies, and farming groups that offer a wealth of resources for small-scale chicken farmers. How can you tap into these resources in your area? Let's find out.

Your County Extension Office

In the US, every county in every state has an extension office, which is an extension of the agriculture-based university in that state. The extension agent or agents work directly with specialists at the university to answer questions, solve problems, and disseminate the most up-to-date information available about farming and agricultural practices, as well as gardening, pest control, soil testing, and more. Because there is an extension office in each county, the agent will have information and resources that are focused on factors that are important for your area.

You can visit your county's extension office—you can find the address, phone number, and hours online—to pick up brochures and pamphlets that have been written and printed for hobby chicken farmers like you. You can even chat with an agent to have your questions answered free of charge.

USDA

The United States Department of Agriculture (USDA) has an abundance of information available to help suburban chicken farmers, including the latest information on avian influenza and biosecurity, feed and nutritional requirements, and best practices to promote good health. Even if there is not a USDA office in your community, you can still access their resources via their website.

Local 4-H Clubs

Our children were already members of a 4-H club when we decided to get backyard chickens. Several families in our 4-H club were experienced in raising chickens and were happy to offer advice and suggestions. No matter how odd the question, there was someone who had an answer for us. Any issue we encountered, there was someone who had dealt with the same problem in the past.

Even if your children are not interested in joining 4-H, you can still reach out to a 4-H parent and pick their brains about suburban chickens.

Local Chicken Hobbyist Groups

As the popularity of suburban chicken farming grows, so has the number of chicken hobbyist groups. It is very likely that there are a few groups in your area. Search Facebook groups to see what is in your area.

You can also connect with local chicken groups by asking your friends, family, neighbors, and coworkers if they know of any groups. Even if they don't have direct information to share with you, they may say something like, "Let me ask my sister-in-law. She's had chickens for years," or "I'll bet my neighbor knows of a group. He is always talking about his hens."

Next time you take a stroll through your neighborhood, make note of houses that have chickens in the backyard. If you see someone outside tending to their flock, you can strike up a friendly conversation and find out if there is a group you can join.

Like my experience with 4-H families, having a network of other suburban chicken farmers in your own community gives you an ideal source of support and information.

Online Resources

The internet is a fast and easy way to get information about suburban chicken farming any time of the day or night. A quick Google search can answer many of the questions you have. A word of warning about online resources, though. As with anything you read online, check the credibility of the source to make sure that you are getting accurate information. Good, solid information is posted by hatcheries, poultry organizations, and government agencies, but there are also websites that have been created by people who don't have the same caliber of credentials or experience. Their information may not be wrong, but you should verify it before you make any decisions based on it.

The other concern with online resources for suburban chicken farmers is location. Look to see where the website is based and for what geographic region the information is geared. I have landed on websites that offer advice for backyard chicken raising in England and Hawaii and Ontario. I live in none of those areas; therefore, the information might not be relevant to me. In the Google search bar, just add "in (your state)" after each search to make sure you are accessing resources for your region.

Blogs and Forums

During your online searches, you will stumble upon quite a few blogs and forums. Many of these are well-written, informative, and entertaining. They can be a source of inspiration and clever ideas too. What I enjoy about blogs is that they are written by actual backyard chicken owners, so their articles are relatable. It's always validating to see that others are facing the same challenges or have the same questions as you.

Forums offer a way for people to ask questions and seek answers to help them with various situations they encounter. Often, however, you have to wade through the responses to find any with actionable information. Some people take the time to respond with thorough, engaging, and useful information, but many do not.

Like websites, seek out blogs and forums that target your location so the information you obtain works for you.

Books

When we first began exploring the possibility of getting chickens, we invested in a few guidebooks. We still refer to them years later. They are like owners' manuals for chickens, providing information on all aspects of raising chickens. There is no shortage of books on Amazon or at your local bookstore that will offer valuable information to help set you on the path the chicken ownership.

Magazines

While *Backyard Poultry* is the largest and best-known magazine for chicken hobbyists, it is not the only one. Next time you are at your local farm and feed store, take a moment to look at the magazine rack. You will see several glossy magazines with beautiful cover photos of handsome chickens. These are fun to read and full of entertaining articles; however, they are not a how-to guide for folks just getting started with suburban chickens. Unlike guidebooks, they do not offer step-by-step instructions for newcomers to the chicken world.

YouTube

There is an entertainment element to YouTube videos, and then there is an educational element. I have found several helpful videos on YouTube that have guided us through chicken conundrums, like building a chicken run, cleaning a tiny chick with pasty butt, and evaluating the freshness of eggs. You may not want to rely on YouTube to guide you through the entire process of preparing for and raising chicks, but you will find answers to many of your questions, as well as visual instructions to help you tackle specific tasks.

Conferences and Workshops

Various groups and organizations sponsor workshops and conferences around the country for backyard chicken farmers. These events bring in experts who can provide you with the most up-to-date information about a wide range of poultry topics, including disease prevention and local ordinances. When you attend a conference or workshop, you will have an opportunity to ask questions and get expert advice to help you with your suburban flock. In addition, you will be in a room filled with other small-scale poultry enthusiasts who will be happy to share their experiences and solutions with you. You will find that chicken people really enjoy talking about their hens. Workshops offer great networking opportunities.

Online Classes and Webinars

Several online learning platforms, and even some universities, offer online classes, e-courses, and webinars to help people get started raising chickens in their suburban backyards. These courses are comprehensive and informative and will offer you complete and accurate assistance in starting a small suburban flock. Most of these courses are video based; however, some are more interactive. Costs vary greatly. If you are interested in gathering the information you need via an online class, e-course, or webinar, do your homework so you can select one that will be a great resource for you and will fit into your budget.

Conclusion

Throughout the 15 chapters of this book, we have discussed the housing needs of suburban chickens, how to select the right breed, where you can purchase chicks, and how to set up a brooder box. We went through the nutritional requirements of chicks, pullets, mature hens, and various types of commercial chicken feed on the market. We learned about ways to keep your hens safe from predators, the pros and cons of free-ranging, and how to protect them from the harsh winter cold. You learned how to find the laws, regulations, and ordinances in your area that dictate the size and restrictions on your suburban flock. We explained egg production, the pros and cons of keeping a rooster, and how to incubate your eggs to hatch your own chicks. Lastly, you discovered ways to keep your chickens, your family, and your community disease-free and healthy using biosecurity, sustainable practices, and good coop hygiene.

This is a lot of information to absorb at once, and new chicken owners can become overwhelmed with it all. My best advice is to relax and enjoy the process. Most of the hard work takes place before you even bring your baby chicks home. By the time the chicks are ready to move

into their backyard coop, you will be much more comfortable with your new role as chicken farmer. And by the time the first eggs start appearing in the nesting boxes, you will have your chicken routine down. Once this happens, you will have become quite knowledgeable about raising chickens.

Chickens are low maintenance. They don't require a large time commitment each day. As long as they have food, water, and shelter, they are content. They will reward your efforts with fresh eggs that are tastier and more nutritious than the eggs you would buy at the supermarket. Raising your own food is empowering, and if you have children, it is educational too.

Of course, you will want to spend time with your little flock to learn their individual personalities and observe their complex pecking order. They are also comical, personable, and friendly. They will give you hours of cheap entertainment. Technically, chickens are livestock, but on a small scale and in a suburban setting, they are more like pets. I have never had a nameless hen, but I have had a Chickaletta, Zsa-Zsa, Tiny, Lady Greyington, Big Mama, and Hennifer Anniston.

Raising chickens in your suburban backyard is a fun, interesting, and beneficial hobby. As Kristin Harrell of Iowa explains, "Being a chicken owner is far more rewarding than I thought it would be." Elizabeth Sorby of Washington seconds that sentiment. "Chickens are delightful."

www.ingramcontent.com/pod-product-compliance
Lightning Source LLC
Chambersburg PA
CBHW070919120626
46546CB00001B/324